Distillation Topical Conference 2012

Topical Conference at the 2012 AIChE Spring Meeting and 8th Global Congress on Process Safety

Houston, Texas, USA
1-5 April 2012

ISBN: 978-1-62276-568-3

Printed from e-media with permission by:

Curran Associates, Inc.
57 Morehouse Lane
Red Hook, NY 12571

Printed by Curran Associates, Inc. (2012)

AIChE
3 Park Avenue
New York, NY 10016-5991

Phone: (203) 702-7660
Fax: (203) 775-5177

www.aiche.org

Additional copies of this publication are available from:

Curran Associates, Inc.
57 Morehouse Lane
Red Hook, NY 12571 USA
Phone: 845-758-0400
Fax: 845-758-2634
Email: curran@proceedings.com
Web: www.proceedings.com

TABLE OF CONTENTS

(30a) AIChE Column Performance Evaluation Guidelines

Ping Xu, UOP LLC, Tonawanda, NY, Daniel R. Summers, Sulzer Chemtech USA, Inc., Tulsa, OK and T.J. Cai, Fractionation Research, Inc., Stillwater, OK

The AIChE has a committee that addresses regularly AIChE's equipment testing procedures. That committee includes a subcommittee that addresses the testing of trayed and packed columns. After three years of work, that subcommittee is ready to release a 150-page manual entitled, "Trayed and Packed Columns – A Guide to Performance Evaluation." This will be the third edition of this manual. This presentation describes the third edition, especially changes with respect to earlier editions.

(30b) Modeling Solvents for Amine-Based CO2 Absorption

Prakash Karpe, Clint P. Aichele and David H. Van Wagener, Sustainability Technologies, ConocoPhillips Company, Bartlesville, OK

Aqueous amine absorption is a mature and proven technology for treating natural gas and refinery fuel gas containing $CO_2$and H_2S. For that reason, it will be a key technology for capturing $CO_2$from flue gas streams for carbon capture and storage. However, the large gas volume and low pressure associated with flue gas streams will result in process units much larger and more expensive than conventional acid gas treaters. Optimization of the design of the absorber and regenerator is important for full-scale CO_2capture in order to minimize both capital and operating costs. This presentation focuses on the development of process modeling using rate-based modeling software for amine absorption. The modeling conclusions incorporate appropriate factors to accurately calculate the overall CO_2avoided, and proper technology comparison strategies are discussed. Regeneration energy comparisons are made between the conventional, 30 wt.% MEA system and other solvents of interest.

(30c) Rate Based Modeling - the Developer's Perspective

Nathan A. Hatcher, Optimized Gas Treating, Inc., Buda, TX

20^{+}years of research, development, and commercialization have culminated in the wide acceptance and success of the ProTreat mass transfer rate-based gas treating simulator. As engineers, we tend to focus on the end product without giving much thought to the thoughts and decisions that are required along the journey to reach a goal. A glimpse into the mindset of engineering software development from the developer's perspective is provided. In particular, certain traits are almost pre-requisites for success in this area of chemical engineering.

(30d) Comparison of Rate-based Results Vs Nutter Ring FRI Test Results using ChemSep

Attilio J. Praderio, LNG Technology Development, ConocoPhillips, Houston, TN and Ross Taylor, Clarkson University, Potsdam, NY

ChemSep is a stand-alone single column simulator for distillation, absorption, and extraction operations. It combines the classic equilibrium stage column model with a nonequilibrium (rate-based) column model in a single program. In this work ChemSep was used to simulate tests carried out by FRI on Nutter Rings #2 (selected from FRI Topical Report # 99) in order to compare the results from several rate base simulations against FRI reported field results. Simulated results matched field results but, as with any simulation software, different results can be obtained when inappropriate options are selected for the specific case under study. The effect of the choice of the binary interaction parameters for the selected equation of state, the choice of the mass transfer models, and methods used for estimation of physical properties of the mixtures are demonstrated in the presentation.

(30e) New Strategies for Acid Gas Enrichment

Tofik K. Khanmamedov, TKK COMPANY, Sugar Land, TX

Sulphur plant performance and capacity are greatly affected by feed gas quality. Thus, increasing the H_2S content of Claus plant feeds is needed when plant throughput is limited or the sour gas has too low an H_2S:CO_2ratio for producing satisfactory acid-gas. The key to acid gas enrichment (AGE) is selectivity, and of the generic amines,*N*-methyldiethanolamine (MDEA) is the most selective. The FLEXSORB® family of hindered amines is even more selective; however, these solvent are more expensive than generics. Recently, a new patented family of HIGHSULF™ amine-based desulphurization technologies and strategies was introduced in the literature. Among other applications, the HIGHSULF technology enhances the well-known SCOT or similar processes, uses identical equipment, requires almost no additional capital investment, and produces extremely high quality Claus feed with generic MDEA, or MDEA spiked with promoters. The beauty of HIGHSULF is its simplicity—any existing AGE or SCOT unit can be switched easily to HIGHSULF and back to conventional treating without shutdown. This paper uses the ProTreat™ process simulator to substantiate the technology and to discuss several cases of both the original and extended HIGHSULF process strategies. Attention focuses on using HIGHSULF with MDEA solvents to upgrade low-quality SRU feed and to produce high H_2S gas from the SRU's TGTU amine section. FLEXSORB or other special solvent performance can be had using HIGHSULF with generic MDEA.

(30f) Bibliography of Dr. Ralph Weiland

Mike Resetarits, Simon Chambers and T.J. Cai, Fractionation Research, Inc., Stillwater, OK

Dr. Ralph Weiland's contributions to chemical engineering have been prodigious. Throughout his entire career he has been at the frontier of research in the fields of distillation, absorption and modeling. His company's ProTreat computer program is now globally well accepted for the simulation of acid gas absorption systems. This presentation will highlight certain articles from Dr. Weiland's career.

(55a) Crude Vacuum Column Wash Section - A New Design Approach

Carlos Trompiz, Fluor, Houston, TX

Process simulations using the results of a model developed by Trompiz and Fair at the Separations Research Program (SRP - University of Texas at Austin) for predicting liquid entrainment from sprays can closely reproduce the actual operation of the wash section of crude vacuum distillation columns. The process simulation should use the net wash oil flow rate which reaches the wash bed and not the total wash oil flow rate measured and controlled in the crude vacuum unit operation. This new approach of simulating a crude vacuum column provides the right basis for new designs or revamps involving the wash section of crude vacuum units to prevent coke formation in the wash bed and ensure its proper performance.

(55b) The Impact of Joint Rotation Angle on the Performance of Mellapak N250Y

Michael Basdenand R. Bruce Eldridge, Chemical Engineering, The University of Texas at Austin, Austin, TX

Structured packing elements are typically installed such that the channels of an element are perpendicular to the element above. Previous studies by Owens*et al.*have shown that this leads to significant recirculation and stagnation zones for single phase flows; increased pressure drop was also observed.[1] In this study, different arrangements of Mellapak N250Y were tested using computational fluid dynamics (CFD) to predict the dry pressure drop as a function of the joint rotation and gas flowrate. The computational predictions for the packing elements oriented perpendicular to one another were validated by experimental data.

1. Owens, S. A.; Kossmann, A.; Farone, J.; Eldridge, R. B.; Ketcham, R. A., Flow Field Visualization in Structured Packing Using Real Time X-ray Radiography. *Industrial & Engineering Chemistry Research* **2009,** 48, (7), 3606-3618.

(55c) Testing of MellapakPlus 352.Y At Fractionation Research, Inc

Marc B. Wehrli, Sulzer Chemtech AG, Winterthur, Switzerland and **Mark W. Pilling**, Sulzer Chemtech USA, Tulsa, OK

Sulzer MellapakPlus^{TM}352.Y packing was tested at Fractionation Research, Inc. in 2011. This high performance packing showed excellent results in the FRI test systems. Test results and interesting comparisons with other high performance packings will be discussed and analyzed.

(55d) FRI Test Data and Performance Modeling of Montzpak B1-250MN

Bjoern Kaibel[1], Helmut Jansen[1], Thomas Rietfort[1], Egon Zich[1]and Zarko Olujic[2], (1)Julius Montz GmbH, Hilden, Germany, (2)Process & Energy Laboratory, Delft University of Technology, Delft, Netherlands

Results are presented of the total reflux distillation experiments carried out with Montzpak B1-250MN at FRI, in a 1.22 m ID column using o/p xylene at 0.1 bar and cyclohexane/n-heptane (CH/n-H) system at 0.31 and 1.62 bar, respectively. All experiments were performed using a Montz high turndown liquid distributor, which performed accordingly over a very wide range of specific liquid loads.

As observed, B1-250MN, the flagship of the newest generation of Montz high performance corrugated sheet structured packings, utilizing perforated sheets with characteristic shallow embossed texture in combination with a reduced corrugation inclination angle and a proprietary smooth bend on lower end of corrugations, exhibits a stable and highly efficient performance over the whole range of operating conditions as encountered in FRI total reflux distillation tests. Interesting to mention is that efficiency curve measured with close boiling, o/p-xylene system matches perfectly that measured under same operating pressure at Bayer TS (BTS) using chlorobenzene/ethylbenzene in a 0.59 m ID column. In both cases the same maximum useful capacity was reached, while the pressure drop observed in FRI test was significantly lower than that measured at BTS. With CH/n-H system slightly lower efficiencies were measured, compared to close boiling system. Practically same efficiencies were obtained at vacuum and above atmospheric pressure, with the pressure drop and maximum useful capacity distributed accordingly. In all cases a strikingly good performance (HETP = 0.25 m at pressure drop below 1 mbar/m) was observed at lowest column loads, and a sudden departure in efficiency was observed upon reaching the point of onset of flooding.

Delft model, with full utilisation of the original correlation tor turbulent vapour flow mass transfer coefficient and a physically founded enhancement of the flow direction change related pressure drop proved to be capable of approaching closely measured efficiency and pressure drop both qualitatively and quantitatively.

Paper No. 55e, Performance Comparison of Structured Packing HETP Correlations

A.K. Jammula and J.R. Whiteley
School of Chemical Engineering
Oklahoma State University

T.J. Cai and M.R. Resetarits
Fractionation Research, Inc.
Stillwater, OK

KEYWORDS

distillation structured packing HETP literature correlations

ABSTRACT

This paper presents an analysis and comparison of five popular HETP correlations for structured packing. The performance comparisons were generated using a screened database containing all of the published HETP experimental results for traditional (not high-capacity) structured packing. The database includes FRI test results which have been released to the public. The screening process excluded data points that fall outside the preloading region. The full database consists of 809 points obtained from 8 different sources. After screening, a total of 766 points were retained for performance evaluation of the different HETP correlations. The total number of packings included in the database is 24. Both hydrocarbon and aqueous systems are represented in the database.

HETP prediction performance is documented for the following correlations: Harrison and France - 1989 [1], Billet - 1999 [4], SRP - 1996 [5], Hanley - 2011 [6] and Delft - 2011 [7]. Performance comparisons are provided using several different correlating factors, e.g. packing type, system type, etc. The paper concludes with a brief discussion of the overall findings.

INTRODUCTION

Accurate HETP predictions are required for proper design and operation of packed columns. Numerous researchers have published HETP research data and results for a variety of structured packing types and operating systems. We have collected all of the published data for traditional structured packing tests (high-capacity structured packing excluded). The full database consists of 809 points obtained from 8 different sources. The complete database was screened to remove 43 data points that fall outside the preloading region. The remaining 766 data points were retained to evaluate performance of five HETP prediction correlations. The distribution of retained data points in the correlation evaluation database is shown in Table 1.

Table 1. Traditional Structured Packing HETP Measurements Published in the Open Literature that Lie Within the Preloading Region

Literature Data Sources	Data Points
SRP	302
Prof. Billet	163
TU Delft	34
Koch	39
Norton	90
FRI	102
Sulzer	11
Spiegel	25
Total no. of points	**766**

The retained data cover 24 different packings in columns ranging from 0.1 – 1.2 m in diameter. Column pressure ranged from 0.021 – 4.14 bar. The following binary systems are represented in the database:

C6 (cyclohexane) / C7 (iso-heptane)
n-octane / toluene
CB (chlorobenzene) / EB (ethylbenzene)
o-xylene / p-xylene
methanol / ethanol
EB (ethylbenzene) / styrene
polyglycol / water

Aspen Plus was used to generate physical properties in situations where physical properties were not reported.

CORRELATIONS

One shortcut and four fundamental HETP prediction correlations were evaluated. Each of the correlations is described briefly below.

Shortcut Correlation

Harrison and France - 1989 [1]

For troubleshooting purposes, Harrison proposed a rule of thumb for both random and structured packings. According to Harrison, structured packing HETP is fixed for a fixed crimp height. According to his rule, HETP is 9 inches for ¼ - inch crimp height, 18 inches for ½-inch crimp and 33 inches for 1-inch crimp. This rule in equation form is

$$HETP = \frac{100}{a_p}$$

This rule is only applicable to packings with a 45° corrugation angle.

Fundamental Correlations

SRP – 1996 [5]

The SRP correlation was developed based on the two-resistance approach. The prediction of HETP requires k_g, k_L and a. In the SRP model, the k_g correlation was developed based on a wetted wall column analogy, the k_L correlation was developed based on penetration theory, and the a correlation was developed based on the earlier work of Shi & Mersmann [8]. SRP model requires a packing specific constant F_{SE} and crimp dimensions for the estimation of effective mass transfer area.

$$k_G = 0.054\frac{D_G}{s}\left[\frac{\rho_G s\,(u_{Le} + u_{Ge})}{\mu_G}\right]^{0.8}\left(\frac{\mu_g}{D_g\rho_g}\right)^{0.33}$$

$$k_L = 2\sqrt{\frac{0.9 D_L u_{Le}}{\pi s}}$$

$$\frac{a_e}{a_p} = F_{SE}\frac{29.12\,(We_L Fr_L)^{0.15} s^{0.359}}{Re_L^{0.2}\epsilon^{0.6}(1 - 0.93\cos\gamma)(\sin\theta)^{0.3}}$$

$$HETP = \frac{\ln\lambda}{\lambda - 1}\left(\frac{u_G}{a_e\,k_G} + \lambda\frac{u_L}{a_e\,k_L}\right)$$

Billet - 1999 [4]

Billet proposed an HETP correlation which can be applied up to the flooding region. The correlation was developed based on the penetration theory and requires packing specific constants. The correlation can be applied to both random and structured packings. Packing specific constants for various random and structured packings were reported by Billet.

$$k_G = C_V\frac{1}{(\varepsilon - h_L)^{\frac{1}{2}}}\frac{a^{\frac{1}{2}}}{d_h^{\frac{1}{2}}}D_V\left(\frac{u_V}{a\,\nu_V}\right)^{\frac{3}{4}}\left(\frac{\nu_V}{D_V}\right)^{\frac{1}{3}}$$

$$k_L = c_L 12^{\frac{1}{6}}\sqrt{\frac{u_L D_L}{h_L d_h}}$$

$$\frac{a_{Ph}}{a} = 1.5\,(a\,d_h)^{-0.5}\left(\frac{u_L d_h}{\nu_L}\right)^{-0.2}\left(\frac{u_L^2\,\rho_L\,d_h}{\sigma_L}\right)^{0.75}\left(\frac{u_L^2}{g d_h}\right)^{-0.45}$$

$$HETP = \frac{\ln\lambda}{\lambda - 1}\left(\frac{u_G}{a_{ph}\,k_G} + \lambda\frac{u_L}{a_{ph}\,k_L}\right)$$

Hanley - 2011 [6]

Hanley developed an HETP correlation by utilizing the point HETPs in the column, unlike other correlations where an average value was used for the calculation of HETP. Hanley did not make any assumptions for the development of the k_g, k_L and a correlations. These correlations were developed using the Buckingham pi theorem and the Chilton & Colburn analogy [9]. The correlation does not require packing specific constants. This correlation is limited to the preloading region.

$$k_L = 0.33\, Re_L Sc_L^{\frac{1}{3}} \left(\frac{c_L D_L}{d_e}\right)$$

$$k_G = 0.0084\, Re_V Sc_V^{\frac{1}{3}} \left(\frac{c_V D_V}{d_e}\right) \left(\frac{\cos\theta}{\cos\frac{\pi}{4}}\right)^{-7.15}$$

$$\frac{a_m}{a_d} = 0.539\, Re_V^{0.145} Re_L^{-0.153} We_L^{0.2} Fr_L^{-0.2} \left(\frac{\rho_V}{\rho_L}\right)^{-0.033} \left(\frac{\mu_V}{\mu_L}\right)^{0.090} \left(\frac{\cos\theta}{\cos\frac{\pi}{4}}\right)^{4.078}$$

$$HETP = \frac{G}{a_m}\left(\frac{C_y}{k_y} + \frac{C_x}{k_x}\right)$$

Delft – 2011 [7]

The Delft correlation was first developed in 1999 and has been routinely updated as additional Montz experimental data became available. The Delft vapor phase mass transfer coefficient correlation was developed based on transitional flow and the liquid phase mass transfer coefficient correlation was developed from the penetration theory. The Delft effective mass transfer area correlation is a modified Onda correlation [10]. The Delft correlation does not require any packing specific constants, but it does require packing crimp dimensions for the calculation of HETP.

$$k_G = \sqrt{k_{G,lam}^2 + k_{G,turb}^2}$$

$$k_{G,lam} = \left(\frac{D_G}{d_{hg}}\right) 0.664\, Sc_G^{\frac{1}{3}} \sqrt{Re_{Grv}\frac{d_{hG}}{l_{G,pe}}}$$

$$k_{G,turb} = \left(\frac{D_G}{d_{hG}}\right) \frac{Re_{Grv}\, Sc_G \frac{\xi_{GL}\varphi}{8}}{1 + 12.7\sqrt{\frac{\xi_{GL}\varphi}{8}}\left(Sc_G^{\frac{2}{3}} - 1\right)} \left[1 + \left(\frac{d_{hg}}{l_{G,pe}}\right)^{\frac{2}{3}}\right]$$

$$k_L = 2\sqrt{\frac{D_L u_{Ls}}{\pi\ \varepsilon\, h_L 2s}}$$

$$\frac{a_e}{a_p} = (1 - \Omega)\left\{1 - \exp\left[-1.45\left(\frac{0.075}{\sigma}\right)^{0.75} Re_{LS}^{0.1} Fr_{LS}^{-0.05} We_{LS}^{0.2}\right]\right\}$$

$$HETP = \frac{\ln \lambda}{\lambda - 1} \left(\frac{u_G}{a_e\, k_G} + \lambda \frac{u_L}{a_e\, k_L} \right)$$

RESULTS

Predictive performance, reported as mean absolute relative error (MARE), of each of the five correlations using the SRP experimental data are summarized in Table 2.

Table 2. Summary of Literature Correlation Performance for SRP HETP data

Packing	Harrison Deviation %	SRP Deviation %	Billet Deviation %	Hanley Deviation %	Delft Deviation %
Flexipak 2	8	23	10	26	24
Gempak 2A	21	30	10	28	22
Intalox 2T	33	29	15	14	10
Maxpak	43	22	18	11	8
Montz B1-250.60	13	19	11	45	15
Montz B1-250	24	24	26	17	12
Montz B1-350	15	21	36	13	20
Montz B1-400.60	26	32	20	37	17
Montz B1-400	8	29	34	10	33
Montz BSH-400.60	25	36	15	39	17
Montz BSH-400	7	20	20	10	29

Literature correlation performance for all databases with parity plots will be presented in the AIChE meeting. The full paper will include a discussion of the overall results.

Bibliography

1. Harrison, M., Distillation Column Troubleshooting, Part 2: Packed Columns. Chem. Eng. 1989, 121.
2. Kister, H. Z., Distillation Design. McGraw-Hill: 1992.
3. Lockett, M. J., Easily predict structured-packing HETP. Chem. Eng. Prog., 1998, 94(1), 60–66.
4. Billet, R.; Schultes, M., Prediction of Mass Transfer Columns with Dumped and Arranged Packings: Updated Summary of the Calculation Method of Billet and Schultes. Chemical Engineering Research and Design 1999, 77 (6), 498-504.
5. Rocha, J. A.; Bravo, J. L.; Fair, J. R., Distillation Columns Containing Structured Packings: A Comprehensive Model for Their Performance. 2. Mass-Transfer Model. Industrial & Engineering Chemistry Research 1996, 35 (5), 1660-1667.
6. Hanley, B.; Chen, C.-C., New mass-transfer correlations for packed towers. AIChE Journal 2011.

7. Olujic, Z. In Predicting the Liquid-Side Mass-Transfer Resistance of Structured Packings: State of the Art, AIChE, Chicago, Illinois, 2011; p 71b.
8. SHI , M., Mersmann A., Effective interfacial area in packed columns. German Chemical Engineering 1985, vol. 8 (no2), pp. 87-96.
9. Chilton, T. H.; Colburn, A. P., Mass Transfer (Absorption) Coefficients Prediction from Data on Heat Transfer and Fluid Friction. Industrial & Engineering Chemistry 1934, 26 (11), 1183-1187.
10. Onda, K., Mass Transfer Coefficients Between Gas And Liquid Phases In Packed Columns. Journal of Chemical Engineering of Japan 1968, 1 (1), 56-62.

(55f) Packing Structure Capable of Sub-Centimeter HETP for High Purity Distillation of Low Relative Volatility Compounds

Ward E. Tegrotenhuis, Pacific Northwest National Laboratory, Richland, WA and Michael R. Powell, Thermal & Reaction Systems, Pacific Northwest National Laboratory, Richland, WA, US Virgin Islands

Achieving a high separation factor with mixtures of low relative volatility typically requires a large numbers of stages and very tall columns, even with advanced structured packings. The challenge is exacerbated by vapor and liquid phase dispersion when a high purity product is demanded. A novel microchannel technology is presented that utilizes layered structures to precisely control the mass transfer length-scale in the liquid phase to a fraction of a millimeter, thereby enabling sub-centimeter height of a theoretical plate (HETP). Furthermore, both the liquid and vapor phase flows are laminar, which reduces dispersion and allows high purity to be achieved even with low relative volatility mixtures. Experiments conducted with low concentrations of 1,4-dimethyl cyclohexane and 3-methyl heptane in octane successfully demonstrated HETP values below 0.5 cm for relative volatilities of 1.03 and 1.24, respectively. The lack of HETP dependence on relative volatility is evidence that dispersion is not limiting product purity. Scale-up of the technology is discussed.

() Musings on Management

Frank van Lier, The Lubrizol Corporation, Wickliffe, OH

It really is all about the people!

Hiring good people.

Retaining them.

Adjusting for different personal styles, cultures, encouraging diversity, embracing different perspectives.

Motivating people to work together in successful global teams.

Achieving top notch results that are better than any individual can achieve alone.

Maintaining the focus.

Getting the results.

Management is all about the excitement managers gets from having their people and teams grow and succeed.

(94a) Multichordal Gamma Scans with Quantitative Analysis for Diagnosing Tray Maldistribution

Henry Kister, Process Engineering, Fluor Corp., Aliso viejo, CA

MULTICHORDAL GAMMA SCANS WITH QUANTITATIVE ANALYSIS: KEY TO DIAGNOSING TRAY MALDISTRIBUTION

Henry Z. Kister, Fluor Corp., Aliso Viejo, CA

Large-diameter olefins towers like primary fractionators and C3 Splitters occasionally experience liquid and/or vapor maldistribution in either conventional or high-capacity trays. Such maldistribution often leads to tray efficiencies much lower than design, forcing an increase in reflux and reboil in the tower, and limiting tower capacity. When such maldistribution occurs, it may be difficult to diagnose using conventional troubleshooting techniques such as vendor software, delta P measurement, and conventional single-chord qualitative gamma scanning. Judicious multichordal gamma scanning with quantitative analysis is a very useful tool for diagnosing maldistribution on trays. This paper discusses this technique and describes its application in a couple of case studies.

(94b) A Dying Column, Hidden Damage And The Weeping Trap Syndrome

Sean Hennigan, BP Hull Research and Techonlogy Centre, Saltend, United Kingdom

A deterioration in a petrochemicals column performance offered a challenging puzzle. Investigations were carried-out into the potential root causes, through data analysis and trials, all of which suggested a severe stage loss through the system - this was at odds with the column scan which showed much more restricted damage. Opening the column showed extensive manway loss and panel damage, apparently confirming a simple explanation of poor manway installation. However detailed analysis of other causes led us to highlight a previously underestimated vulnerability with this kind of column named "Weeping Trap Syndrome" or WTS. Cases where the steady state load profiles in the column change severely tray to tray are most susceptible to WTS. Whilst most trays are well loaded others are on the edge of weeping even at design.

Following this theory through, the loss of a few trays can have a significant overall stage loss as the low load front shifts down the column. For most towers the numbers of stages is a threshold function of rate - hit the right load to seal the trays and the stage count is fixed. For a WTS column we postulate an odd feedback mechanism where the number of stages depends on both rate and the degree of separation you achieve. This leads to potentially unworkable startups.

(94c) Effect of Tray Spacing on Tray Performance

Simon Chambers, T.J. Cai, A.N. Vennavelli and A. Y. Ogundeji, Fractionation Research, Inc., Stillwater, OK

At the 2007 AIChE Spring Meeting, FRI presented information on the effect of deck open area and hole size on tray performance. The theme of tray design effects on performance is extended to tray spacing. The impact of three tray spacings on capacity, efficiency, pressure drop and rangeability will be described and illustrated. We will also show that the impact of tray spacing on tray capacity is other than TS**0.5. FRI is currently performing additional research work in air-water and hydrocarbon systems regarding closely-spaced trays.

APPLICATION OF ULTRA HIGH CAPACITY SHELL CONSEP* TRAY FOR DE-BOTTLENECKING OF A CRUDE DISTILLATION UNIT

Kaushik Majumder, Shell Projects & Technology, Bangalore, India
Giuseppe Mosca, Sulzer Chemtech Ltd., Winterthur, Switzerland
Kent Mahon, The New Zealand Refining Co. Ltd., Whangarei, New Zealand

Abstract

The main fractionator of the Crude Distillation Unit (CDU-1) in Whangarei refinery of The New Zealand Refining Company Limited (NZRC) was retrofitted with high capacity internals to increase the unit throughput from 8500 ton/day to 13000 ton/day. Shell ultra high capacity ConSep trays were applied in the most capacity constrained HGO pump-around (mid circulating reflux) section of the column as no other first generation high capacity tray was found adequate to debottleneck this section. By the application of ConSep trays, capex savings of the order of USD 5.5 - 6 million was achieved compared to other conventional debottlenecking options. This was the first application of ConSep trays in a CDU main fractionator and the post revamp test run established the realization of the expected performance.

Introduction

NZRC's Whangarei refinery targeted expanding the refining capacity through 'Point Forward Project' [1]. The project involved increasing the throughput of the CDU-1 from 8500 ton/day to 13000 ton/day, thereby increasing distillate component to downstream processing and generating additional long residue to replace imported long residue for loading the vacuum distiller. Figure-1 shows the simplified process flow diagram of CDU-1.

Shell Global Solutions International (SGSi) carried out the feasibility study for the expansion of CDU-1. Several options were studied to debottleneck the main fractionator:

1. Replacement of the existing column internals with high capacity internals including ConSep tray for the most capacity constrained HGO pump-around section.
2. Installation of a new crude pre-fractionator column to separate off light naphtha and reduce the load to the main fractionator. The capex for this option was found USD 6 million higher than for option 1.
3. Installation of a new heavy end column to recover the Heavy Gas Oil (HGO) dropped into the long residue to offload the main fractionator. The capex for this option was found USD 5.5 million higher than for option 1.

On basis of the comparison of the revamp options, NZRC has decided to proceed with the ConSep tray alternative owing to its lowest capex and most favourable economics.

Shell ConSep Tray Technology

The ConSep tray utilizes the principle of de-entrainment by centrifugal forces to remove the gravitational limitation of jet flood. Separation of the entrained liquid before entering the next tray allows very high vapour velocities to be achieved in the column. The ConSep tray combines the features of a contacting deck and a separator deck in a single tray. The basic features of the ConSep tray are illustrated in Figure-2. The functioning of the contacting deck, which in fact is a normal tray, is limited by three hydraulic mechanisms: jet flooding, downcomer choking and downcomer backup. The use of separator deck influences all three mechanisms [2]:

1. The jet flooding limit is extended as the entrained liquid is efficiently separated from the vapour to prevent carry-over of liquid to the tray above.
2. The liquid entering the main downcomer is largely coming from the separator deck, where it is well degassed. As a result, the downcomer liquid handling capacity is substantially increased.
3. To eliminate downcomer back-up limitation, the separator deck is designed with low pressure drop swirl tubes. The contacting deck is also designed with relatively high open area.

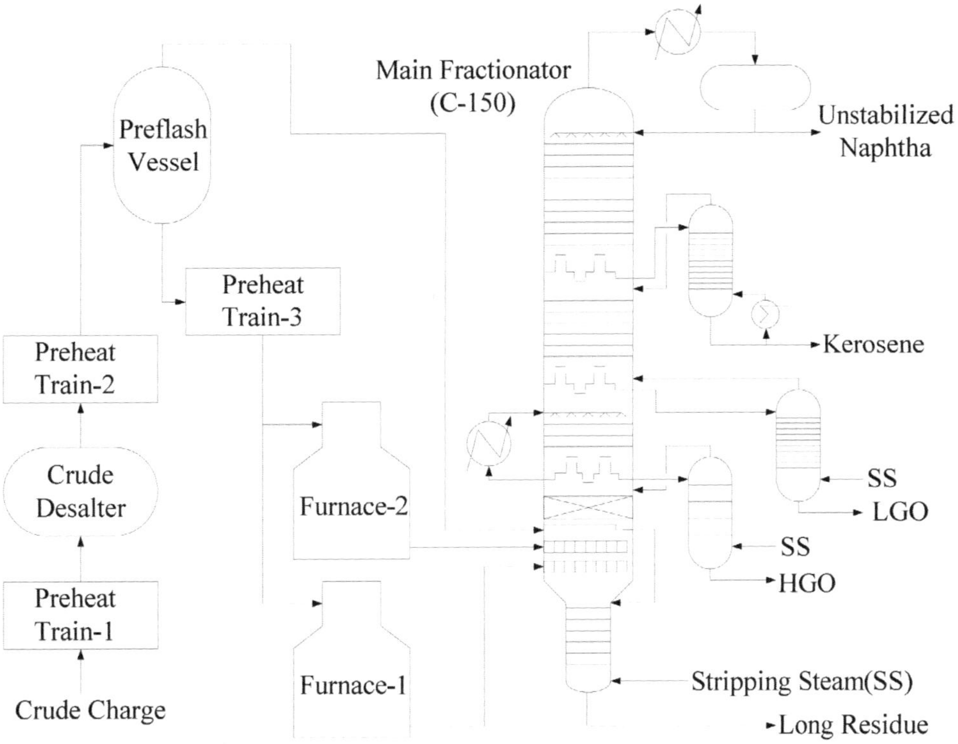

Figure-1: Simplified process flow diagram of CDU-1

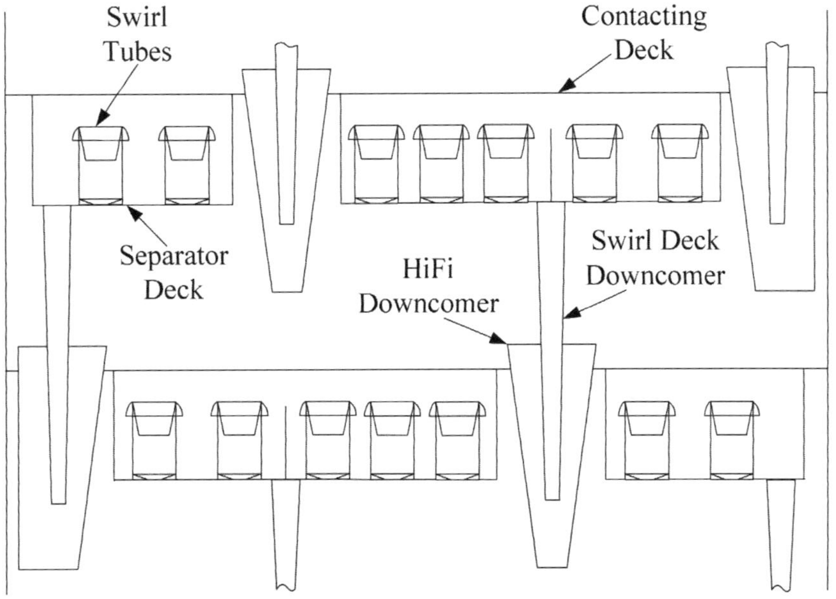

Figure-2: Schematic drawing of ConSep tray

Figure-3 illustrates the expected capacity gain of the ConSep tray over conventional trays and packing [3].

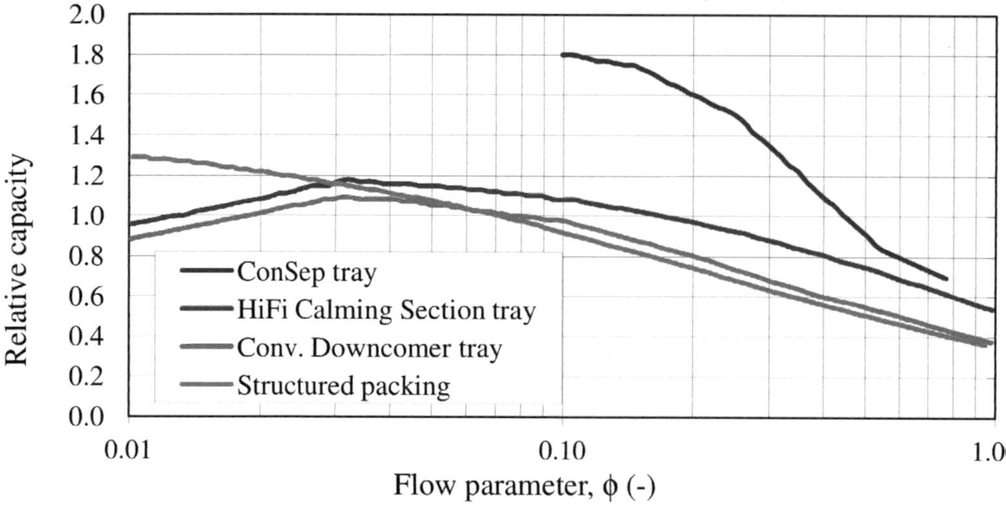

Figure-3: Comparison of column internal capacities

The flow parameter (ϕ) is defined as $\phi = \frac{L}{V}\sqrt{\frac{\rho_V}{\rho_L}}$, where $\frac{L}{V}$ represents the liquid-to-vapour mass flow ratio and $\frac{\rho_V}{\rho_L}$ represents the ratio of vapour-to-liquid density.

Typically the ConSep tray is capable of offering 40-50% capacity advantage over a wide range of first generation high capacity trays. In most revamps, the column retrofitted with the ConSep tray gets limited by other factors such as availability of feed and/or constraints on auxiliary equipment e.g. reboiler, condenser, pumps etc. even before the full potential of the ConSep tray is realized. Table-1 shows some of the applications of the ConSep tray along with benefits achieved and constraints faced.

Table-1: Overview of ConSep tray applications

Location	Diameter	Application	Installed	Capacity Increase, %
UK	1.9	NGL Debutanizer	1995	22 [(1)]
Australia	1.9	FCCU Debutanizer	1996	30 [(2)]
Germany	2.2	HCU Main Fractionator	1999	50
Australia	1.7	NGL Debutanizer	1999	15 [(1)]
Singapore	2.5	FCCU Debutanizer	2000	20 [(1)]
Japan	2.1	FCC Debutanizer	2006	10
Sweden	2.0	FCC Debutanizer	2006	20
Sweden	1.0	C3/C4 splitter	2007	50
U.S.A.	2.3	C3/C4 splitter	2006	12
Canada	1.0	Depropanizer	2007	20
New Zealand	4.6	Crude Distillation	2009	22 [(2)]
Singapore	1.8	PO Drying Column	2011	20 [(2)]
China	3.2	Ethylene Fractionator	2011	Target 25% [(3)]
Japan	2.5	FCC Debutanizer	2013(scheduled)	Target 15% [(3)]

[(1)] limited by reboiler capacity [(2)] limited by feed to column [(3)] no data yet

Modifications of Main Fractionator (C-150)

During the feasibility study, the HGO pump-around section of the column was found severely limiting for the targeted throughput of 13000 ton/day. This section was already fitted with Shell CS** trays (Calming Section trays). Since the first generation of high capacity internals were found inadequate to debottleneck this section, ultra high capacity Shell ConSep trays were selected. The trays were designed to achieve 33% more capacity compared to Shell CS tray.

The HGO pump-around section consisted of 3 contacting trays with a tray spacing of 500 mm. A one-for-one tray replacement with ConSep trays was selected. Figure-4 shows a schematic drawing of HGO pump-around section fitted with ConSep trays. For the remaining sections of the column, the following internals were suggested:

- Stripping section : Shell HiFi*** trays
- Wash section : Mellapak**** Plus 252Y packing
- All other sections : Shell CS trays

As this was the first application of Shell ConSep trays in this service, a detailed study was carried out to address the risks associated with this application and the mitigations were applied in the design. The ConSep trays were manufactured by Sulzer Chemtech. To ensure proper performance of the trays in a relatively new application, very rigorous quality control steps were followed at the manufacturing site and detailed mock-up assembly of tray components was carried out at refinery site prior to installation in the column (Figure-5).

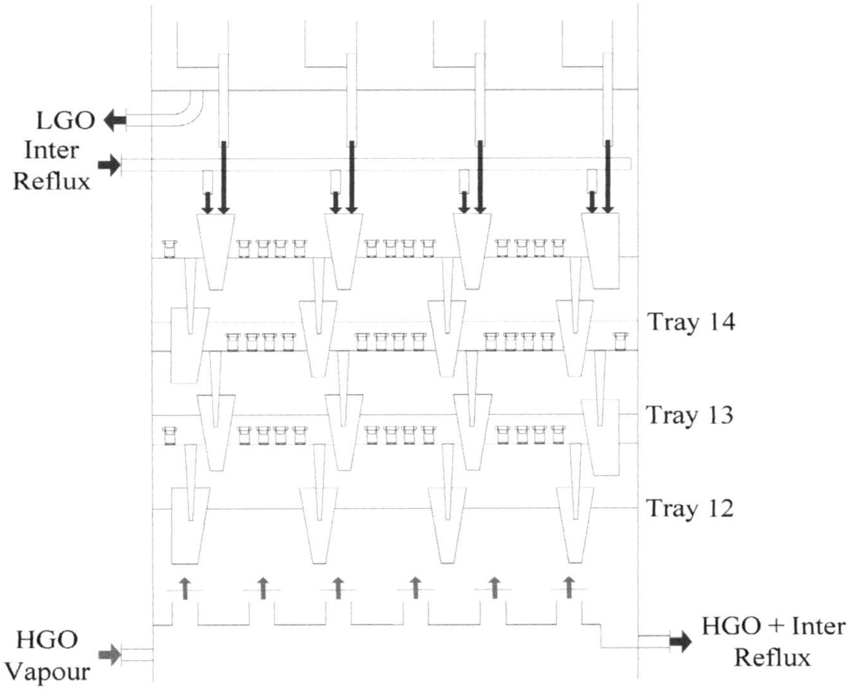

Figure-4: Schematic drawing of HGO pump-around section of C-150

Post Revamp Performance

From conception to implementation, the project took 4 years and the revamp was implemented during a normal planned shutdown in October 2009 without affecting the unit availability. The performance test run of CDU-1 was conducted in September 2010. The crude throughput was maintained at 13000 ton/day, though the crude blend used during the test run was marginally heavier than the one considered for the revamp design. Comparison of the design crude blend and crude blend used during test run is shown in Figure-6.

Figure-5: Inspection/Mockup/Installation of ConSep trays

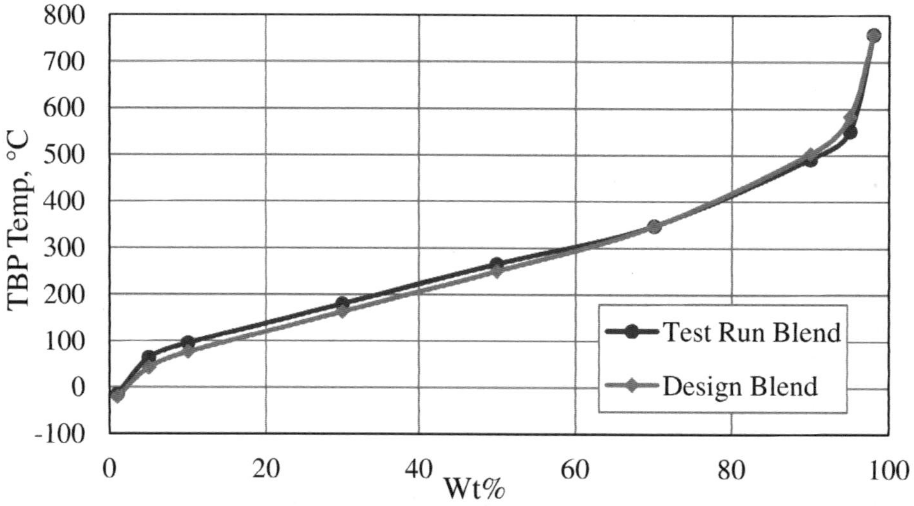

Figure-6: Comparison of design and test run crude TBP

Key design conditions and test run operating conditions of C-150 are presented in Table-2. The product quality is compared in Figure-7(a-e).

Table-2: Key design and test run operating conditions of C-150

Parameters	Unit	Design	Test Run	Parameters	Unit	Design	Test Run
Crude Intake	T/D	13000	13077	Flash Zone Press	barg	1.90	2.24
Naphtha	T/D	4405.0	3516.4	Feed Temp	°C	361.5	345.0
Kerosene	T/D	1498.8	2169.3	Top Temp	°C	180.8	175.7
Light Gas Oil	T/D	1517.0	1116.1	Kero Draw Temp	°C	216.5	223.4
Heavy Gas Oil	T/D	2014.5	1731.9	LGO Draw Temp	°C	244.2	250.1
Long Residue	T/D	3573.3	4271.1	HGO Draw Temp	°C	280.7	278.3
Strip Steam	T/D	106.9	149.5	Flash Zone Temp	°C	343.9	327.3
Top Press	barg	1.65	1.95	Bottom Temp	°C	333.3	322.061

C-150 was simulated for the test run conditions to evaluate hydraulic loading of the ConSep trays. In Table-3, the key performance indicators for the ConSep trays operating under test run conditions are compared with the design conditions. During the test Run, the ConSep trays were operating at 10-15% lower than the design capacity even at design intake of 13000 T/D due to:

- Heavier feed than the design case
- Less preheat recovery of the order of 10-12°C due to limitation in crude preheat train

7a: Naphtha ASTM D86

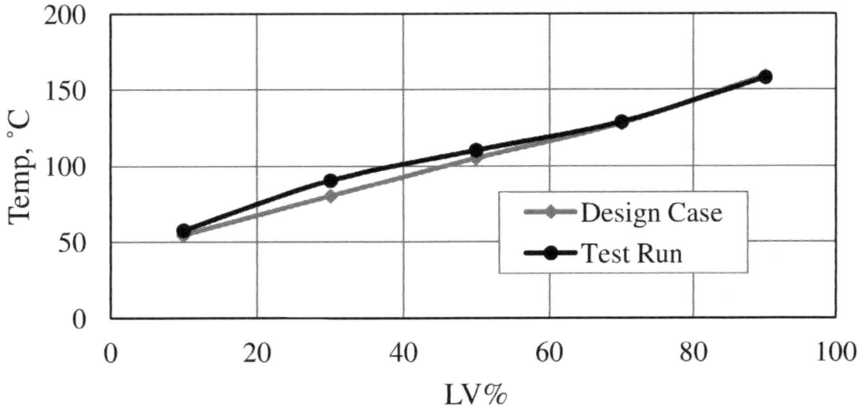

7b: Kerosene ASTM D86

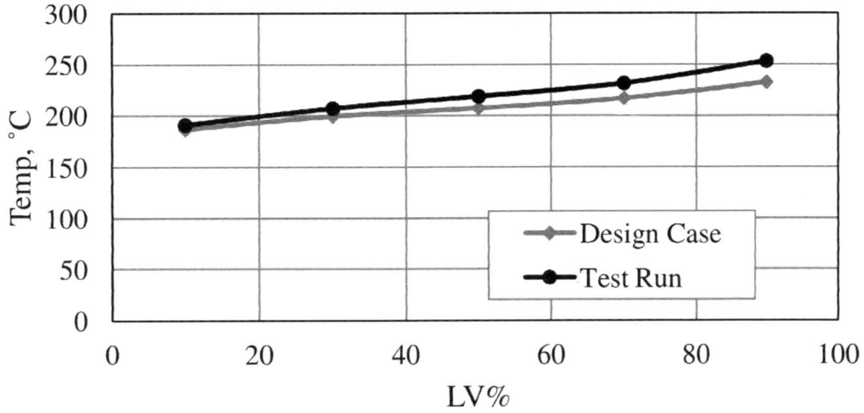

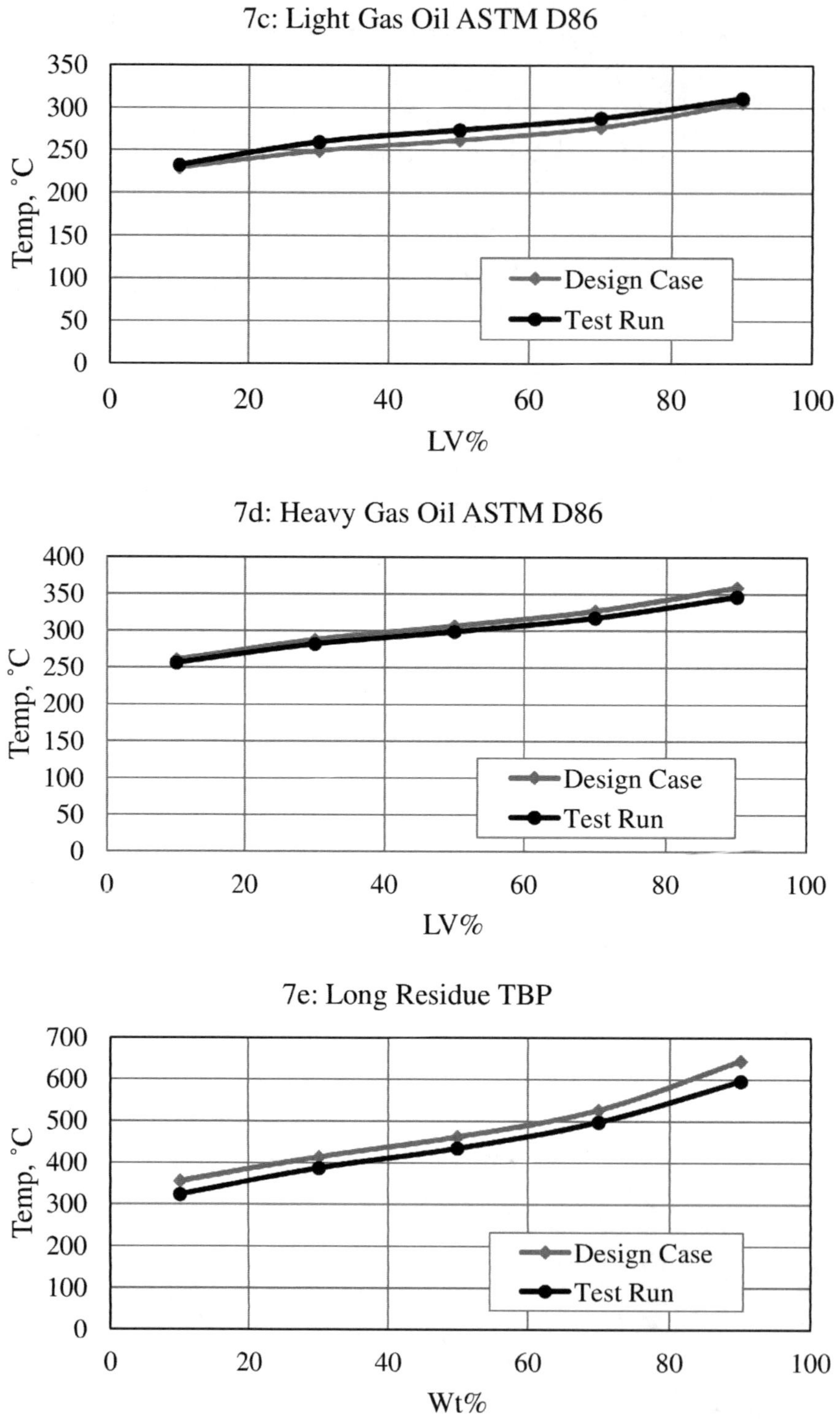

Figure-7(a-e): Comparison of design and test run product quality

Table-3: Key performance indicators for ConSep trays

Parameters	Unit	Design	Test Run
Froth backup / CS Height	%	68	60
Tray pressure drop	mbar	12.3	9.2
Tube flood percent	%	73	60
Flow Parameter	-	0.17	0.19
Overall column load factor	m/s	0.12	0.10
Flooding percent (CS tray)	%	133	112

Based on the review of the operating experience since start-up and the test run results, it could be concluded that the revamp targets of CDU-1 main fractionator (C-150) were achieved. No hydraulic constraint was experienced in achieving the design intake of 13000 ton/day and required product quality was achieved.

Conclusions

The performance of Shell ConSep trays in the HGO pump-around section of CDU-1 main fractionator met the target of capacity enhancement without any draw back compared to the pre-revamp conditions. During the test run, the trays were operating at 10-15% lower than the design capacity even at the design intake of 13000 T/D due to heavier crude feed and lower feed temperature. However, the built-in capacity margin enabled stable operation for the trays at much above the capacity limit of the first generation of high capacity trays.

The options to debottleneck columns already equipped with the first generation of high capacity trays are limited. Shell ConSep trays provide a very attractive solution to such cases. In this revamp project, use of only 3 Shell ConSep trays in the most capacity constrained section of the column made it possible to retrofit the existing column and turned the capex option more attractive over the other debottlenecking options.

References

1. 'Refinery expansion means NZ more self reliant' media release by NZRC dated 16 July 2010
2. Wilkinson, P.M., De Villiers, W.E., Mosca, G., Tonon, L., 'Achieve Challenging Targets in Propylene Yield using Ultra System Fractionation Trays', ERTC 2006.
3. De Villiers, W.E., Bravo, J.L., Wilkinson, P.M., Summers, D.R., 'Further advances in light hydrocarbon fractionation', PTQ Summer 2004

(*) Shell ConSep is a Shell trademark
(**) Shell CS a Shell trademark
(***) Shell HiFi a Shell trademark
(****) Mellapak is a Sulzer-Chemtech trademark

(94e) Flow Pattern Optimization of a Sieve Plate Extraction Column Using CFD Simulation and PIV Measurement

Changchun Duan[1], **Bing Wang**[2], Chunjiang Liu[1] and Xigang Yuan[1], (1)State Key Laboratory of Chemical Engineering, Tianjin University, (2)Chemical Engineering, New Jersey Institute of Technology, Newark, NJ

Present work deals with flow pattern optimization of the continuous phase in a sieve plate extraction column using both computational fluid dynamics (CFD) simulation and particle image velocimetry (PIV) measurement. Firstly, flow pattern of a sieve plate extraction column (SPEC) with traditional configuration was simulated and it was found that there was a very large reverse flow region between every two plates. Then, flow pattern was optimized with the objective of minimizing reverse flow area since the reverse flow can decrease the mass transfer efficiency. Step by step, internal structure of the column was modified to minimize the reverse flow area and with the modification, flow pattern of the continuous phase was optimized. The strategies consist of inclining downcomers, adding diversion baffles, slotting the downcomers and diversion baffles, adjusting the number and size of slots and fixing small baffles laterally at the top slots on downcomers. In the course of evolution, five generations (11 types) of SPEC with different internal configuration were proposed and flow patterns in these columns were investigated. Finally, an optimal flow pattern nearly no reverse flow region was obtained. PIV experiments of three types of columns in 11 were carried out and it was found that the results of simulations and experiments are in good agreement. Besides, the mass transfer performances of all columns were investigated by CFD simulation and it was found that mass transfer efficiency was improved as the flow pattern was optimized. The flow pattern optimization strategies proposed here can be extended to the design of other process equipment such as distillation and reaction.

Keywords: flow pattern optimization, extraction column, CFD, PIV

(94f) Pressure Drop on Gas Tangential Velocity in Rotating Zigzag Bed

Yumin Li, College of Chemical Engineering & Materials Science, zhejiang University of Technology, Hangzhou, China

The pressure drop across Higee is one of the most important characteristics, and the liquid flow gives rise to intriguing and unusual effects on the pressure drop. The pressure drop of the wet Higee can be substantially lower than that of the dry one. This trend is quite different from the conventional packed and trayed tower. And the unusual effects also occur in the rotating zigzag bed(RZB) as a novel type of Higee. The pressure drop of RZB was studied using two methods of pressure drop model based on the gas tangential velocity and empirical correlation. Experiments were carried out in a RZB of 214mm inner-diameter, 486mm outer-diameter and 104mm axial-height rotor with air-water system. The radial distribution of gas tangential velocity in the RZB rotor was obtained. And the liquid flow reduces the gas tangential velocity. The gas tangential velocity was correlated by an empirical equation, on which the centrifugal pressure drop is predicted by the pressure drop model. Thereby, the frictional and momentum pressure drop are also predicted by the model. As a result, the phenomenon that the wet rotor pressure drop is lower than the dry rotor pressure drop is reasonably explained. The variation of the overall pressure drop of RZB is similar to that of the rotor pressure drop, and it is correlated by an empirical equation with good agreement.

(94g) Design of Distillation Columns for Ternary Heterogeneous Azeotropic Mixtures by Geometrical Methods

Cesar Augusto Sánchez Correa[1], **Miguel Ángel Gómez García**[2] and Gerardo Rodríguez Niño[1], (1)Chemical and Environmental Engineering Department, National University of Colombia, Bogotá, Colombia, (2)Chemical Engineering Department, National University of Colombia, Manizales, Colombia

The distillation columns for the fractionation mixtures with regions of partial miscibility allowed the possibility of coexistence of two o more liquid phases inside or in an external decanter. This feature of the distillation columns with heterogeneous mixtures implies some advantages because of two issues associated with the decanter: a) enables cross a thermodynamic separatrix leading to products located in different distillation regions; b) when differences in solubility are notaries enables high purity distillation products in the section where the decanter is localized.

The basic design of a heterogeneous distillation column consists in developing the following tasks: 1) locate the product composition region and verify the separability condition; 2) investigate the conditions of minimum energy consumption (minimum reflux calculation); 3) locate the feeding stages and calculating the number of theoretical stages for a finite reflux (higher than the minimum). Shortcut methods used for the basic design include the knowledge associated with the following geometric elements [1, 2, and 3]: a) the *reversible distillation trajectories* (RDT); b) the distillation borders at infinite reflux; c) the *Pitchfork distillation boundary (PDB); d) the structure of vapor-liquid-liquid equilibrium.*

In this work new tools are present for the basic design of a heterogeneous distillation column. The target is: a) present a continuation method for tracking of the RDT associated with the feeds and the products; b) present a continuation technique for tracking of the PDB; c) use the RDT and the PDB in association with the other geometric elements for the exploring of the conceptual design problem. Three examples of design are illustrated: dehydration of acetic acid with n-amyl acetate, synthesis of separation sequence for mixtures of water + n-pentanol + n-amyl acetate and a review of the dehydration of ethanol with benzene.

We conclude that there are some common and complicated situations in the design of azeotropic heterogeneous distillations: 1) the minimum reflux can be controlled by a tangent pinch point; 2) the range of allowable values for the reflux is very narrow; 3) the minimum reflux not coincide with a infinite number of stages; 4) the admissible specifications in the condenser-decanter depend of the vapor line extension and localization of the thermodynamic separatrix.

Key Words: Heterogeneous azeotropic distillation; reversible distillation; Pitchfork distillation boundary; continuation methods

References

[1]. Lucia, A.; Amale, A. y Taylor, R. "Distillation Pinch Points and More". *Computers & Chemical Engineering*, vol. 32, 2008, pp. 1342-1364.

[2]. Petlyuk, F. B.; Danilov, R. Yu. y Serafimov, L. A. "Trees of Reversible Distillation Trajectories and Structure of Trajectory Bundles for Sections of Adiabatic Columns". *Theoretical Foundations of Chemical Engineering*, vol. 42, No. 6, 2008, pp. 795-804.

[3]. Królikowski, A. R.; Królikowski, L. J. y Wasylkiewicz, S, K. "Distillation profiles in ternary heterogeneous mixtures with distillation boundaries". Chemical Engineering Research and Design, vol 89, July 2011, pp. 879-893

(106a) Synthesis of PVA Based Membranes Modified by Benzimidazole Group for Direct Methanol Fuel Cells

Pratima Gajbhiye, Chemical Engineering, Indian Institute of Technology, Kanpur, State of Uttar Pradesh, India, Jayant K. Singh, Chemical Engineering, Indian Institute of Technology, Kanpur, Ka, India and Anil Kumar, Department of Chemical Engineering, Indian Institute of Technology Kanpur, Kanpur, India

Abstract

In recent years, there has been a great effort to look for alternate membrane material and Polybenzimidazole (PBI) membranes have been synthesized where the conducting moiety of this polymer is benzimidazolium groups. The monomers used to synthesize these polymers are bis conjugated amines and diacids are expensive. In this work, we have synthesized grafted polymers of PVA as membrane materials where the grafts are either Poly Acrylic Acid, Poly maleic anhydride (PMA) or Poly Methacrylic acid (PMAA) and these materials were denoted by PVA-g-PAA, PVA-g-PMA, and PVA-g-PMAA. We then synthesized C1, C3 & C4 compounds (formula given at the bottom of Table S1) having bis conjugated amines whereas compound C2 having only one conjugated amine group. One of the steps in synthesizing C1 compound is nitration of phenyl ring, where there is a possibility of formation of two isomers out of which only one of them has strictly bis conjugated amines and can be polymerized with Isophthalic acid giving the formation of PBI. The other byproduct formed is an impurity and is minimized by carrying out the nitration step at low temperature but this cannot be completely removed from the product which is a mixture. Table S1 presents the summary of our experimental results on characterization of membrane, In the C1 to C4 compounds and in some cases (membranes 9, 10, 11), Sulfocuccinic acid were added to the casting polymer solution as additional crosslinking agent in order to achieve higher conductivity. We added the polymer, thus formed, to the casting solution, we first compared the results of the Fuel cell formed by adding C1 (C3 and C4) compound alone and heating at 180°C for some time. Table S1 has two sets of results, one on characterization (columns 1, 2, 3,4,5,6, and 7) of membranes while the other on Fuel cell (columns 8 to 11). For the latter columns 10 and 11 gave results on characterization on Fuel cell while columns 8 and 9 gave results when the Fuel cell was run on control mode for higher efficiency.

Experiments on membrane pore sizes are given in column 1 which lies between 8.88 to 9.52nm and they all are ultra filtration type. The values of the pore size are mostly governed by the nature of the graft chains, e.g. membranes 1,3,4,7 and 12 consist of PAA grafts and were found to have similar pore sizes. Similarly membranes 2,5,8,9 were of the similar pore size because the PVA have the same grafts of PMA. The next important thing is to determine the wall charges of these membranes and they were estimated by experimentally determining the rejection of Chromic (VI) acid solution through these membranes. Subsequently, a space charge model was proposed for the flow of the solution through their pores and a semi analytical solution was derived for the overall problem of separation of chromic acid solution. We observed that the high surface charge was because of the benzimidazolium groups present on the surface and they ranged from minimum of 465 to maximum of 525mV. The thermal degradation (TGA) experiments were carried out which showed that, through the modification of the polymer by benzimidazole groups, lead to stable membrane and the degradation temperature varies from 150°C for 10% weight loss and for approximately 80% weight reduction, a minimum of 300°C to 700°C.

The twelve membranes reported in Table S1 were first characterized completely in terms of pore size, wall charge, conductivity, IEC and thermal degradation. In fuel cell applications, higher temperatures give large current density and it is observed that some of the membranes (rows 1, 2, 5, 6, 7, 8 and 12) could easily be operated upto 150°C without the membrane degrading. However in our case, we insisted that the membrane of direct methanol fuel cell were operated at 90°C for which most of the fuel cell results for Nafion 117 are available. The catalysts used on the anode and cathode side were Pt-Ru and Pt loaded on carbon respectively, which were synthesized by following procedure given in literature. We find that in each case, the power density and the current density were higher and for some of our results (membranes 1,6,12) were atleast two to three times higher than that of Nafion 117 as seen in column 10 and 11.

It was recognized that the membrane material has hydrophilic PVA matrix and hydrophobic benzimidazolium groups at the surface formed by reaction of the C1 to C4 compounds with grafts (see Fig.S-1). When the reaction occurs (confirmed by IR and NMR), the hydrophobic benzimidazolium are formed which tends to separate as confirmed by the SEM results. In view of this we assumed that the pores of the membrane would have a thin hydrophobic layer with positive charge walls. This is confirmed by the contact angle measurements where the minimum contact angle is 51° and varied upto 89°. Even though most of the polymer matrix is PVA with its grafts, when the ions move through the pores (giving rise to current in Fuel cell), the pore walls get highly charged (as measured by column 2 of Table S1). Some of the membranes have Sulfo succinic acid SSA as part of the matrix (e.g. the membrane 9 to 11) which would tend to give sulfonate groups inside the PVA matrix as well as at the phase boundaries, this way giving rise to bipolar nature of the membrane (benzimidazolium group being strictly cationic in nature). In addition to this, some of the compounds such as C1 and C2 compounds have bonded sulfonate groups which would give the positive and negative charge separated by the molecular distance, once again producing a bipolar nature to the membrane. In remainder of the cases the charges would be induced on the phase boundaries as seen in Fig.S-1. The conductivity observed in our membranes is very high and could be attributed to the special bipolar structure of our polymers.

Table S1. Fuel cell and Membrane Characterization

Compound	Membrane	1	2	3	4	5	6	7	8	9	10	11

			MWCO (pore size, nm)	Wall Charge	Conductivity (S/cm)	IEC	Thermal Degradation temp (°C) 10% - 80%		Water Uptake	Contact Angle	Average Voltage Gain (V)	Average time cycle (min)	Power Density (mW/cm)	Current Density (mA/cm)
C1	1	PVA-g-PAA-C1	9.1	488	0.04	2.7	199.732	268.70	7.5	72	0.523	55.75	50.47	74.88
	2	PVA-g-PMA-C1	9.35	465	1.23×10^{-2}	1.8	142.659	447.6	10.691	74.917	0.568	42.896	0.305	3.81
	3	PVA-g-PAA-modified	9.52	488	1.51×10^{-2}	2.12	115.123	706.53 (28.93%)	13.665	75.581	0.778	69.85	37.37	64.44
C2	4	PVA-g-PAA-C2	9.52	525	6.59×10^{-2}	2.38	119.558	691.93	13.4	51.2	0.704	39.83	23.0020	50.55
	5	PVA-g-PMA-C2	9.36	519	1.04×10^{-2}	2.17	181.59	562.348	14.9	50.24	0.77	161.93	34.72	62.11
	6	PVA-g-PMAA-C2	8.99	482	1.16×10^{-2}	2.18	201.055	472.568	13.8	51.8	0.6066	147.81	76.72	92.33
C3	7	PVA-g-PAA-C3	9.0	512	2.51×10^{-2}	2.23	204.908	453.03	8.624	70.368	0.6883	82.55	36.397	75
	8	PVA-g-PMA-C3	9.5	510	3.51×10^{-2}	2.27	166.819	505.293	13.513	62.944	0.726	37.557	36.11	64.381
	9	PVA-g-PAA-SSA-C3	9.36	497	1.71×10^{-2}	2.51	154.914	813.769 (24.32%)	10.5	54.8	0.8096	99.423	23.45	57.778
	10	PVA-g-PMA-SSA-C3	9.83	480	2.28×10^{-2}	2.46	115.12	428.713	8.5	76.2	0.663	224.22	28.561	58.358
	11	PVA-g-PMAA-SSA-C3	8.88	525	3.66×10^{-2}	1.85	139.772	681.715 (30.78%)	7.2	89.2	0.77	18.36	20.832	48.11
C4	12	PVA-g-PAA-C4	9.5	495	2.56×10^{-2}	1.62	189.87	451.86	8.927	59.132	0.704	74.19	41.1	71
	13	Nafion 117			0.1								21.01	48.33

Compound C1, [Bis-3-Amino-4-{3-(triethylammoniumsulfonato) phenylamino}-phenyl sulfone] hydrochloride

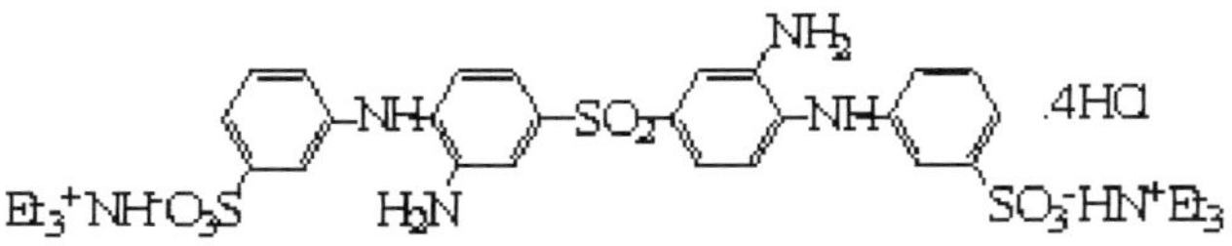

Compound C2, 3-Amino -4-[3-(triethyl ammonium sulfonato) phenyl amino]

phenylene hydrochloride

THE GRAPHIC ON THIS PAGE WAS NOT AVAILABLE ON THE E-MEDIA VERSION OF THIS PAPER.

Compound C3, Bis -4 [1, 2 diamino phenyl]-phenyl sulfone) hydrochloride

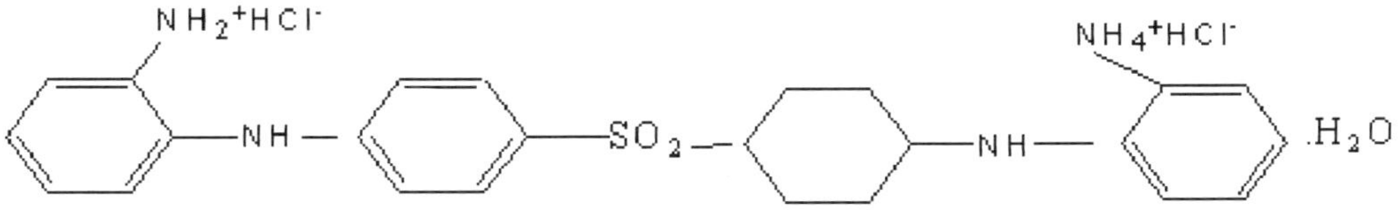

Compound C4, 1, 4 [1, 2 diamino phenyl] Phenylene hydrochloride

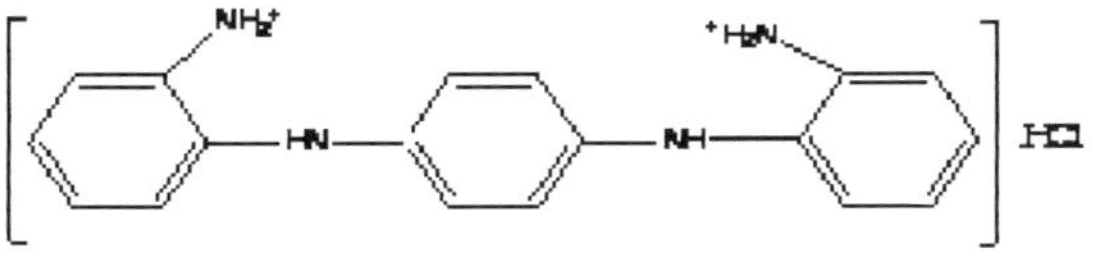

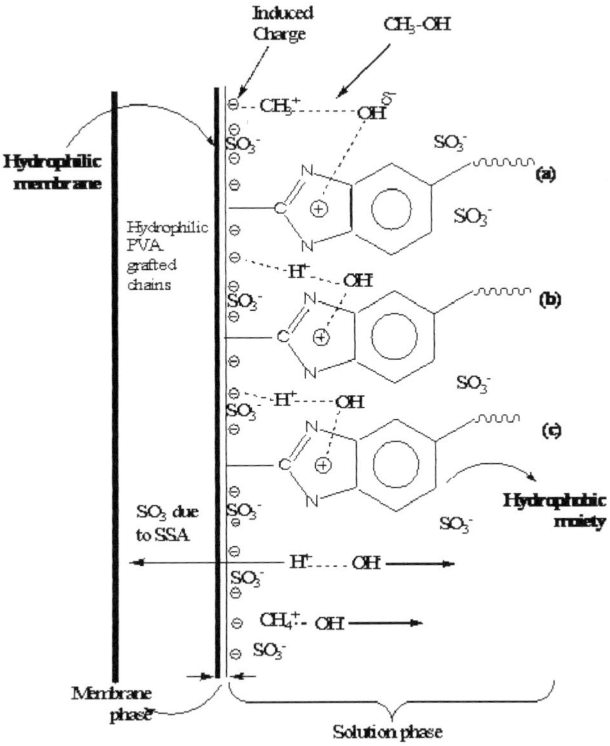

Figure S-1. The Bipolar nature of the membrane surface

(106b) Operation and Controls of Operating Direct Methanol Fuel Cells with PVA Based Membranes Modified by Benzimidazolium Groups

Pratima Gajbhiye, Chemical Engineering, Indian Institute of Technology, Kanpur, State of Uttar Pradesh, India, Pranjal Joshi, Chemical Engg, IIT Kanpur, Kanpur, India, Jayant K. Singh, Chemical Engineering, Indian Institute of Technology, Kanpur, Ka, India and A. Kumar, Department of Chemical Engineering, Indian Institute of Technology Kanpur, Kanpur, India

Operation and Control of Direct Methanol Fuel cell with PVA based membranes modified with benzimidazolium groups

Pratima Gajbhiye, Pranjal Joshi, Jayant K.Singh, Anil Kumar

Department of Chemical Engineering, Indian Institute of Technology, Kanpur 208016, India

Abstract

In recent years, in an effort to find alternate membrane material (other than Nafion 117), Polybenzimidazole (PBI) membranes have been synthesized where the conducting moiety of this polymer is benzimidazolium groups. The monomers used to synthesize these polymers are bis conjugated amines and diacids are expensive. In this work, we have synthesized grafted polymers of PVA as membrane materials where the grafts are either Poly Acrylic Acid, Poly maleic anhydride (PMA) or Poly Methacrylic acid (PMAA) and these materials were denoted by PVA-g-PAA, PVA-g-PMA, and PVA-g-PMAA. We then synthesized C1, C3 & C4 compounds (formula given at the bottom of Table S1) having bis conjugated amines whereas compound C2 having only one conjugated amine group. One of the steps in synthesizing C1 compound is nitration of phenyl ring, where there is a possibility of formation of two isomers out of which only one of them has strictly bis conjugated amines and can be polymerized with Isophthalic acid giving the formation of PBI. The other byproduct formed is an impurity and is minimized by carrying out the nitration step at low temperature but this cannot be completely removed from the product which is a mixture. Table S1 presents the summary of our experimental results on characterization of membrane, In the C1 to C4 compounds and in some cases (membranes 9, 10, 11), Sulfocuccinic acid were added to the casting polymer solution as additional crosslinking agent in order to achieve higher conductivity. We added the polymer, thus formed, to the casting solution, we first compared the results of the Fuel cell formed by adding C1 (C3 and C4) compound alone and heating at 180°C for some time. Table S1 has two sets of results, one on characterization (columns 1, 2, 3,4,5,6, and 7) of membranes while the other on Fuel cell (columns 8 to 11). For the latter columns 10 and 11 gave results on characterization on Fuel cell while columns 8 and 9 gave results when the Fuel cell was run on control mode for higher efficiency.

Experiments on membrane pore sizes are given in column 1 which lies between 8.88 to 9.52nm and they all are ultra filtration type. The values of the pore size are mostly governed by the nature of the graft chains, e.g. membranes 1,3,4,7 and 12 consist of PAA grafts and were found to have similar pore sizes. Similarly membranes 2,5,8,9 were of the similar pore size because the PVA have the same grafts of PMA. The next important thing is to determine the wall charges of these membranes and they were estimated by experimentally determining the rejection of Chromic (VI) acid solution through these membranes. Subsequently, a space charge model was proposed for the flow of the solution through their pores and a semi analytical solution was derived for the overall problem of separation of chromic acid solution. We observed that the high surface charge was because of the benzimidazolium groups present on the surface and they ranged from minimum of 465 to maximum of 525mV. The thermal degradation (TGA) experiments were carried out which showed that, through the modification of the polymer by benzimidazole groups, lead to stable membrane and the degradation temperature varies from 150°C for 10% weight loss and for approximately 80% weight reduction, a minimum of 300°C to 700°C.

The twelve membranes reported in Table S1 were first characterized completely in terms of pore size, wall charge, conductivity, IEC and thermal degradation. In fuel cell applications, higher temperatures give large current density and it is observed that some of the membranes (rows 1, 2, 5, 6, 7, 8 and 12) could easily be operated upto 150°C without the membrane degrading. However in our case, we insisted that the membrane of direct methanol fuel cell were operated at 90°C for which most of the fuel cell results for Nafion 117 are available. The catalysts used on the anode and cathode side were Pt-Ru and Pt loaded on carbon respectively, which were synthesized by following procedure given in literature. We find that in each case, the power density and the current density were higher and for some of our results (membranes 1,6,12) were atleast two to three times higher than that of Nafion 117 as seen in column 10 and 11. In normal Fuel cell operations, there is decay in the cathode catalyst which is further increased by the methanol crossover. The voltage of the Fuel cell increases, the average volatage gained is reported in column 8 and it is found that the voltage gain depends on the nature of the membrane which is not the same for all of them. In column 9, we have measured the average time cycle and we found that some of the membranes (membranes 6, 7, 9, 10) gave extremely good results.

It was recognized that the membrane material has hydrophilic PVA matrix and hydrophobic benzimidazolium groups at the surface formed by reaction of the C1 to C4 compounds with grafts (see Fig.S-1). When the reaction occurs (confirmed by IR and NMR), the hydrophobic benzimidazolium are formed which tends to separate as confirmed by the SEM results. In view of this we assumed that the pores of the membrane would have a thin hydrophobic layer with positive charge walls. This is confirmed by the contact angle measurements where the minimum contact angle is 51° and varied upto 89°. Even though most of the polymer matrix is PVA with its grafts, when the ions move through the pores (giving rise to current in Fuel cell), the pore walls get highly charged (as measured by column 2 of Table S1). Some of the membranes have Sulfo succinic acid SSA as part of the matrix (e.g. the membrane 9 to 11) which would tend to give sulfonate groups inside the PVA matrix as well as at the phase boundaries, this way giving rise to bipolar nature of the membrane (benzimidazolium group being strictly cationic in nature). In addition to this, some of the compounds such as C1 and C2 compounds have bonded sulfonate groups which would give the positive and negative charge separated by the molecular distance, once again producing a bipolar nature to the membrane. In remainder of the cases the charges would be induced on the phase boundaries as seen in Fig.S-1. The conductivity observed in our membranes is very high and could be attributed to the special bipolar structure of our polymers.

Table S1. Fuel cell and Membrane Characterization

Compound	Membrane		1	2	3	4	5		6	7	8	9	10	11
			MWCO (pore size, nm)	**Wall Charge**	**Conductivity (S/cm)**	**IEC**	**Thermal Degradation temp (°C) 10% - 80%**		**Water Uptake**	**Contact Angle**	**Average Voltage Gain (V)**	**Average time cycle (min)**	**Power Density (mW/ cm)**	**Current Density (mA/ cm)**
C1	1	PVA-g-PAA-C1	9.1	488	0.04	2.7	199.732	268.70	7.5	72	0.523	55.75	50.47	74.88
	2	PVA-g-PMA-C1	9.35	465	1.23×10^{-2}	1.8	142.659	447.6	10.691	74.917	0.568	42.896	0.305	3.81
	3	PVA-g-PAA-modified	9.52	488	1.51×10^{-2}	2.12	115.123	706.53 (28.93%)	13.665	75.581	0.778	69.85	37.37	64.44
C2	4	PVA-g-PAA-C2	9.52	525	6.59×10^{-2}	2.38	119.558	691.93	13.4	51.2	0.704	39.83	23.0020	50.55
	5	PVA-g-PMA-C2	9.36	519	1.04×10^{-2}	2.17	181.59	562.348	14.9	50.24	0.77	161.93	34.72	62.11
	6	PVA-g-PMAA-C2	8.99	482	1.16×10^{-2}	2.18	201.055	472.568	13.8	51.8	0.6066	147.81	76.72	92.33
C3	7	PVA-g-PAA-C3	9.0	512	2.51×10^{-2}	2.23	204.908	453.03	8.624	70.368	0.6883	82.55	36.397	75
	8	PVA-g-PMA-C3	9.5	510	3.51×10^{-2}	2.27	166.819	505.293	13.513	62.944	0.726	37.557	36.11	64.381
	9	PVA-g-PAA-SSA-C3	9.36	497	1.71×10^{-2}	2.51	154.914	813.769 (24.32%)	10.5	54.8	0.8096	99.423	23.45	57.778
	10	PVA-g-PMA-SSA-C3	9.83	480	2.28×10^{-2}	2.46	115.12	428.713	8.5	76.2	0.663	224.22	28.561	58.358
	11	PVA-g-PMAA-	8.88	525	3.66×10^{-2}	1.85	139.772	681.715	7.2	89.2	0.77	18.36	20.832	48.11

		SSA-C3						(30.78%)						
C4	12	PVA-g-PAA-C4	9.5	495	2.56×10^{-2}	1.62	189.87	451.86	8.927	59.132	0.704	74.19	41.1	71
	13	Nafion 117			0.1								21.01	48.33

Compound C1, [Bis-3-Amino-4-{3-(triethylammoniumsulfonato) phenylamino}-phenyl sulfone] hydrochloride

Compound C2, 3-Amino -4-[3-(triethyl ammonium sulfonato) phenyl amino] phenylene hydrochloride

Compound C3, Bis -4 [1, 2 diamino phenyl]-phenyl sulfone) hydrochloride

Compound C4, 1, 4 [1, 2 diamino phenyl] Phenylene hydrochloride

THE GRAPHICS ON THIS PAGE WERE NOT AVAILABLE ON THE E-MEDIA VERSION OF THIS PAPER.

Figure S-1. The bipolar nature of PVA-g-PAA-BASPAPS membrane which reduces the methanol crossover due to the molecular dissociation. In (a), the imidazolium group is shown to separate out from the PVA moiety of the chain into a separate phase. In (b), the water is shown to complex with the SO and the imidazolium group. As a result of this, interaction in (d) and (e), the water and the methanol is shown to split into anion and cation and the anion permeates through the membrane giving bipolar effect

(106e) Numerical Investigation in Heat Conduction Effect on Heat Exchange for Mini Cross-Flow Heat Exchanger (MCHE)

Hongyue Duan[1], Wei Mu[1], **Xigang Yuan**[1] and Lingai Luo[2], (1)State Key Laboratory of Chemical Engineering, Tianjin University, Tianjin, China, (2)LOCIE Chambery, Savoie University, Savoie, France

In a traditional heat exchanger, the effect of heat transfer within the wall material driven by the temperature gradient in axial direction of the fluid flow is usually very small because the area of the cross section of the tube wall is much smaller than that of the tube for heat exchange. However, in a mini cross-flow heat exchanger(MCHE), where the heat conducting medium is typically a cubic-shaped block of metal and the hot and cold streams are put into heat exchange via two sets of channels in the block, the heat conduction in the wall material in the direction parallel to the fluid flow may have a significant effect on the performance of the heat exchanger.

In this paper, a 2-channel-layer MCHE was simulated and a periodic boundary condition was applied so that the results could be applicable to entire MCHE. The proposed model and the mesh grid size were proved by the good agreement of the simulated overall heat transfer coefficient with that found in the experimental reported in the literature[1]. Then the effect of heat conduction on heat transfer performance in a MCHE was studied numerically. The simulation results suggest that the performance of the MCHE can be improved by increasing the wall thickness between the adjoining channels in the same layer or by decreasing the wall thickness between the adjacent layers. The heat conduction effect can also be weakened by optimizing the material conductivity or by increasing the diameter of the channels. New correlations were developed based on the simulated results to predict the value of the overall heat transfer coefficient as a function of the operating parameter such as Re in the channel and the design parameters of the MCHE such as the diameter of the channel, wall thickness and thermal conductivity.

[1]Luo L A, Fan Y L, Zhang W W*et al.*Integration of constructal distributors to a mini crossflow heat exchanger and their assembly configuration optimization[J]. Chemical Engineering Science, 2007, 62 (13): 3605-3619.

(106g) The Effects of Out Gassing Through Cylindrical Walls on the Transport Coefficients

Luma. H. Yaseen[1],**Nada B. Nakkash**[2]and M. N. Latif[1], (1)Al-Nahrain University, Baghdad, Iraq, (2)Chemical Engineering, Al-Nahrain University, Baghdad, Iraq

Abstract

This work concerns with studying the influence of blowing on friction factor and mass transfer coefficient when the flow is normal to a tube bank with outgassing porous cylinders. The friction factor and the mass transfer coefficients are measured for a range of Reynold's number and blowing rate and expressed in terms of various dimensionless groups.

The test rig consists of a rectangular duct (20.3 × 11.3 cm), five rows of three porous tubes in staggered arrangements fixed across the duct with 1.5pitches (both longitudinal and transversal). The outside and the inside diameters of tubes were 25.4 and 21.4 mm, respectively.

The friction factor measurements was studied for Reynold's number range of

$3000 < Re < 17{,}000$and blowing rate up to 4×10^{-3}. The results show that the blowing intothe boundary layer for the cylinders increases the overall friction factor from 10%to 50% depending on the blowing rate. The mass transfer coefficient was measured for Reynold's number range of $5000 < Re < 16{,}000$ and blowing rate of $10^{-5} < m/G_\infty < 10^{-3}$.

A new definition of the blowing rate was considered and expressed by the mass transfer driving force. Two successful formulas were obtained to account the outgassing effect on the mass transfer coefficient. The mass transfer coefficient for the outgassing boundary layer was obtained first by determining the non-blowing coefficient and multiplying it with the correction factor .

The first model was modified to account the density variation associated with the boundary layer change:

$$St_m Sc^{2/3} = \frac{0.51\,Re^{-0.5}}{(1+\beta_m)^{4.7}}\left(\frac{\rho_s}{\rho_\infty}\right)^{0.4}$$

With the coefficient of determination R2 = 0.93.

The second model depends on the flat plate correction factor:

$$St_m Sc^{2/3} = 0.5\,Re^{-0.5}\left(\frac{Ln(1+\beta_m)}{\beta_m}\right)^{9.94}$$

With the coefficient of determination R2 = 0.92.

These two models are found to be in agreement with the previous literature and a reduction of about 80% in the mass transfer coefficient has been obtained due to the outgassing.

(106h) Down the Hubbert Curve

Ian Sutton, Sutton Technical Books, Houston, TX

Down the Hubbert Curve

In March 1956 M. King Hubbert, who was at that time Chief Geologist with the Shell Development Company published a seminal paper with the title "Nuclear Energy and the Fossil Fuels" (Hubbert 1956). In this lengthy and authoritative paper, which was presented before the American Petroleum Institute in San Antonio, Texas, Dr. Hubbert showed how key fossil fuel resources in the United States, particularly coal and oil, were of finite size and that their output would reach a maximum, and then decline. Dr. King's forecasts for the production of oil in the United States were accurate; he predicted that oil production would peak in the early 1970s, and that is what has happened. (The widely-used term "Peak Oil" is frequently accredited to Dr. Colin Campbell (2001)).

In the fifty years since the publication of Dr. King's paper great advances have been made in finding and extracting oil from what were considered to be marginal fields. But Dr. King's overall conclusion is still valid. Production of oil in the United States remains below the 1970's peak. Although Dr. Hubbert's analysis was confined to the United States, subsequent analysts applied his insights to other parts of the world. For example, the book *Twilight in the Desert: The Coming Saudi Oil Shock and the World Economy* (Simmons 2005) by the late Matthew Simmons achieved wide recognition by showing that the concept of peak oil most likely applies to reserves in Saudi Arabia, and other nations such as Iran and Kuwait.

The purpose of this paper is to examine how oil production and consumption have changed in the last five years, and to examine whether or not the concept of Peak Oil remains valid. Key to this discussion is an understanding of the concept of Energy Returned on Energy Invested (EROEI). It is concluded that the concept of Peak Oil, as first postulated by Dr. King, remains valid and current, and the production of oil peaked in the period 2005-2011 and is now set to move down the right hand side of the Hubbert Curve.

The title of this session is *Challenges in Instituting Energy Savings Project in the Present Market Conditions*. If the production of oil and has indeed already passed a peak then the challenge is not to do with "instituting" energy savings; those savings are going to occur whether we like it or not. The challenge will be to determine how the process industries can restructure themselves given that the energy resources available (both as direct energy and as petrochemical feedstock) will be in decline and that there will be intense demand for those same resources from other areas of the economy.

(106j) Statistical Decision Making for In-Service Inspection of Underground Waste Storage Tanks

Emily Mitchell, Emory University, Atlanta, GA and **Stephen P. Harris**, Savannah River National Laboratory, Aiken, SC

Pitting is a localized form of corrosion that occurs on passivated metal surfaces exposed to aggressive chemical environments. The pitting process involves the breakdown of the passive oxide film and subsequent anodic dissolution of the metal at a local site, while the remainder of the surface remains protected by the film.

Deep pits may result in lowering the maximum level of waste that made be stored in the tank and eventually lead to loss of confinement. Pit depths, established by ultrasonic testing, are subject to variables that cause uncertainty, including the repeatability of metric location, time between measurements, different transducers, and different inspectors.

The generalized extreme value (GEV) distribution is fit to the data gathered from ultrasonic testing and statistical projections are made for the full tank wall interior. A comparison is made with results from other well known statistical distributions. The estimated parameters are used in statistical tolerance limits for bounding pit depth in a waste tank.

The tolerance limits give an indication of extreme values of pit depth for the current condition of the tank walls. The upper tolerance limit for pit depth will then be compared to action levels for pitting. If the upper tolerance limit for pit depth exceeds the action level, more frequent examinations or increased tank surface area inspections are conducted.

(106k) Thermal Inbreathing Requirements of Low Pressure Storage Tanks At Elevated Temperatures

Donald E. Brooks, Pressure Relief Analysis, Lloyd's Register Energy Americas, Inc., Houston, TX

ISO 28300 (API 2000) "Petroleum, petrochemical and natural gas industries - Venting of atmospheric and low pressure storage tanks" provides guidance on determining the required thermal inbreathing flow rates to protect atmospheric and low pressure storage tanks against vacuum due to weather changes (e.g. a decrease in atmospheric temperature)." For heated, un-insulated tanks with vapor space temperatures above 58.9 °C (120 "F), ISO 28300 recommends an engineering review of thermal inbreathing requirements.

This paper describes one approach to evaluating thermal inbreathing requirements for un-insulated storage tanks with vapor space temperatures exceeding 58.8 °C (120 "F)." The tank is assumed to be exposed to the cooling effects of a sudden rain. A simplified unsteady state heat transfer analysis is performed to determine the tank wall and vapor space temperatures versus time. Initial gas mass and pressure are assumed to remain constant in the tank. Gas volume in the tank is calculated as a function of vapor space temperature using the ideal gas law. The rate of volume change is calculated for each time increment. The rate of volume change goes through a maximum value. This maximum value is selected as the thermal inbreathing requirement of the tank.

The calculation methodology described in this paper can be programmed in a spreadsheet for ease of use.

Dynamic Operational Risk Assessment with Bayesian Networks

Shubharthi Barua, Xiaodan Gao, M. Sam Mannan
Mary Kay O'Connor Process Safety Center
Texas A&M University, Jack E. Brown Engineering Bldg., 3122 TAMU, Room 244,
College Station, TX 77843
shubharthi.barua@tamu.edu, xiaodan_gao08@mail.che.tamu.edu, mannan@tamu.edu

Extended Abstract Prepared for the Poster Presentation at the
8th Global Congress on Process Safety
Houston, TX
April 1 – 4, 2012

UNPUBLISHED

8TH GLOBAL CONGRESS ON PROCESS SAFETY

Dynamic Operational Risk Assessment with Bayesian Networks

Shubharthi Barua, Xiaodan Gao, M. Sam Mannan

Mary Kay O'Connor Process Safety Center
Texas A&M University, Jack E. Brown Engineering Bldg., 3122 TAMU Room 244, College Station, TX-77843
shubharthi.barua@tamu.edu, xiaodan_gao08@mail.che.tamu.edu, mannan@tamu.edu

Keywords: risk assessment, Bayesian network, dynamic Bayesian network, fault tree, event tree

Abstract

Oil/gas and petrochemical plants are complicated and dynamic in nature. Dynamic characteristics include ageing of equipment/components, season changes, stochastic processes, operator response times, inspection and testing time intervals, and timing of safety system operations, all of which are time dependent criteria that can influence dynamic processes. Conventional risk assessment methodologies can quantify dynamic changes in processes with limited capacity. Several methods are proposed for dynamic risk assessment such as C-DET, D-DET, DORA, etc., but all of them have deficiencies of some kind. Bayesian networks, a graphical inference technique, have some advantageous characteristics for risk analysis over other techniques. This research focuses on providing a framework for dynamic risk analysis based on Bayesian networks. A case study will be provided to illustrate this model.

1.0 Introduction

Oil/gas, chemical, and petrochemical process plants consist of numerous equipment and unit operations, thousands of control loops, and exhibit dynamic behavior. Different hazards and several types of risk are associated with these process facilities. For example, large process areas have the potential to stock a large amount of hazardous chemicals, which can result in the process area being highly congested. It is reported that process risk, i.e., risk in process facilities due to fires and explosions, greatly contributes to the total risk of offshore installations. Also, reactivity, toxic, and mechanical hazards have a significant effect on the total risk of these facilities. Therefore, it is very important to identify hazards, perform risk assessments, and take proper initiatives to minimize/remove hazards and risks; else a catastrophic accident may result.

From case histories, it has been observed that catastrophic accidents have a significant effect on people, environment, and society. Catastrophic accidents such as the Flixborough disaster, the Bhopal incident, and the Piper Alpha disaster caused fatalities and unbearable economic loss. In the past few years, several disastrous accidents in refineries, power plants, and offshore platforms have significantly affected people's perception, and consequently raised concern to emphasize process safety. A series of explosions in 2005 resulting from insufficient process safety and lack of risk reduction measures at the BP Texas City refinery accident resulted in 15 fatalities, 180 injuries, and $1.5 billion in losses [1]. The Kleen Energy power plant explosion in

Connecticut in 2010 is considered to be one of the worst industrial explosions in the USA [2]. In 2010, a fire and explosion resulting from a blowout at the Macondo well resulted in 11 deaths and 17 injuries. Also the continuous spill from the wellhead for 87 days had disastrous effects on the environment and wildlife surrounding the Gulf of Mexico. It has been reported that most of the accidents could be prevented or their severity could have been minimized by effective risk assessment and process safety management. But existing risk assessment methodologies cannot properly quantify various time-dependent effects. Therefore, the objective of this research is to develop a risk assessment methodology with the capability to capture variations due to different time-dependent effects.

2.0 Literature Review

N. Siu [3] summarized available methodologies for risk assessment. For example, the fault tree and event tree methods are two most common methodologies for risk assessment. But both techniques have limited ability to address different time-dependent effects such as equipment or component ageing, stochastic processes, operator response times, and inspection and testing time intervals. N. Siu [3] also described other methodology limitations while applying dynamic system analyses. Therefore, for dynamic systems, conventional techniques are not able to provide accurate results because of their inadequate ability to describe the variation of operational risks as time-dependent deviations, or the changes occurring in the process. Continuous event trees [4], discrete event trees [5], and Markov theory [6] methods have been proposed for dynamic reliability and safety analysis. But all of the proposed techniques have some deficiencies. For example, the Markov theory based model does not consider the effect of inspection on system-state transitions. X. Yang et al. [1] proposed the Dynamic Operational Risk Assessment (DORA) technique, which is able to take inspection and testing time intervals and repair time into account for evaluating operational failure.

Recently, Bayesian networks are receiving wide acceptance in reliability assessment, fault diagnosis, safety, risk, and maintenance analysis. Several authors have [7,8] compared some commonly used methods for accident analysis with Bayesian networks, and described several advantages of Bayesian networks over these methods. N. Khakzad et al. [2] illustrated parallelism between Bayesian networks and fault trees, and discussed the potential of Bayesian networks as a tool for process safety and accident analysis. H. Pasman and W. Rogers [9] described an approach to incorporate Bayesian networks into layer of protection analysis to make it more effective. These studies demonstrate the potential advantages of Bayesian networks application in risk assessment. This research provides a methodology for dynamic risk assessment based on Bayesian networks.

3.0 Methodology and Expected Result

3.01 Bayesian Networks

A Bayesian network, which is a probabilistic graphical model, describes causal influence relations among variables. A set of random variables and their conditional dependencies are represented by Bayesian networks via a directed acyclic graph (DAG). It has the ability to represent dependency among events clearly, accommodate multi mode and continuous random variables, and incorporate information i.e. generic, system specific and expert judgment to support optimum decision making.

A simple Bayesian network is shown in Figure 1. In a binary network, variables are represented by nodes and causal relationships among different nodes are shown by an arc. Conditional probability tables or defined probabilistic relations among nodes represent how one variable is linked with one or many variables. The nodes that influence another variable and have unconditional probability are called parent or root nodes. Nodes that are conditionally dependent on their direct parents are called intermediate nodes. The top node is defined as a leaf node, which is similar to the top event of a fault tree and initiating event for an event tree.

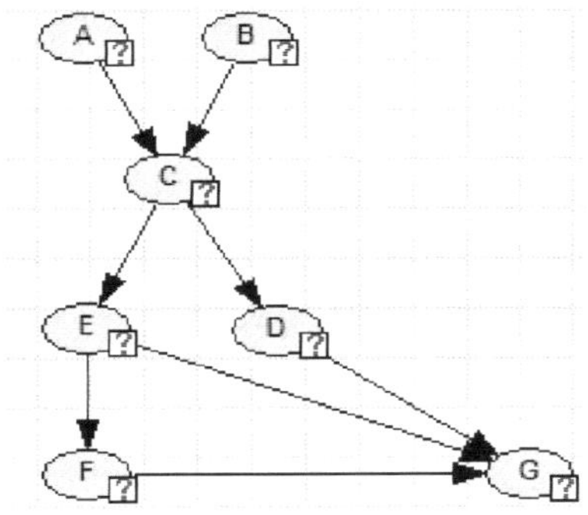

Figure 1: A simple Bayesian network

Let $N = (G, P)$ be a Bayesian network, where

$G = (V, E)$ is a directed acyclic graph; V (random variables) represents nodes; and E represents edges between pairs of nodes of DAG.

P represents probability distribution over V and $V = \{X_1, X_2, \ldots, X_n\}$ is either discrete or continuous random variables [10]. These random variables are assigned to the nodes and the edges.

Bayesian networks can be represented by the joint probability distribution P(V);

$$P(V) = \sum_{X \in V} P\{X|pa(X)\} = P(X_1, X_2, \ldots, X_N) = \sum_{i=1}^{n} P\{X_i|pa(X_i)\}$$

Here $pa(X_i)$ = parent nodes of X_i.

A main advantage of the application of Bayesian networks in risk analysis is the ability to update prior data using Bayes' theorem by incorporating new information as demonstrated by N. Khakzad et al. [2]. Also, Bayesian networks have an advantage of handling different types of uncertainty.

3.02 Dynamic Bayesian Networks

A general Bayesian network is a static model, i.e., the joint probability distribution is usually a representation of a fixed point or an interval of time [11]. A dynamic Bayesian network is a Bayesian network that has the ability to describe the evolution of joint probability distribution over time. Dean and Kanazawa [12] proposed modeling time in a discrete manner to represent the progression of time in a dynamic Bayesian network. In a dynamic Bayesian network, arcs links node/nodes from previous time slice to the node/nodes of the next time slice to represent temporal dependencies among them.

3.03 Methodology

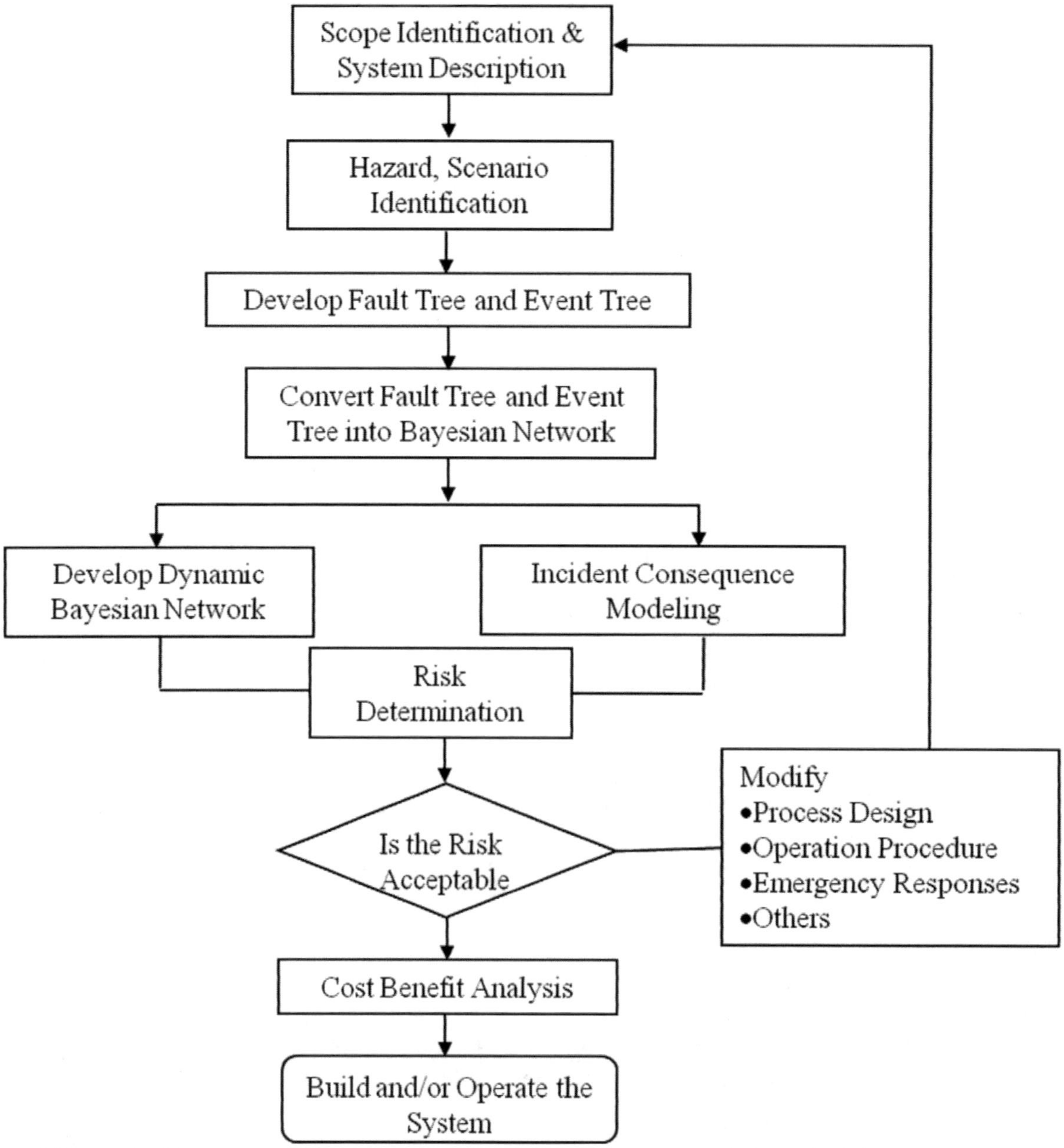

Figure 2: Framework for the dynamic Bayesian network based risk assessment

The objective of this research is to propose a dynamic risk assessment method based on dynamic Bayesian networks. According to the framework, a fault tree and event tree are developed for a system. The objective for following step is to convert the developed fault and event tree into a Bayesian network. Bobbio et al. [13] provided a methodology for conversion of fault trees into Bayesian networks. N. Khakzad et al. [2] showed equivalency between fault trees and Bayesian networks, and described the advantages of Bayesian networks for process safety analysis. Bearfield and March [14] proposed another methodology for generalizing event trees into Bayesian networks. The method has the ability to generalize simple event trees into Bayesian networks in a relatively easy manner, but the method becomes complicated for large event trees. Therefore, in the first phase, a methodology is proposed to convert fault trees and event trees into a static Bayesian network. In the second phase, a dynamic Bayesian network is developed from the static Bayesian network. The dynamic Bayesian network provides probability of a system at different time-slices for different states of the system. The probabilities are then combined with consequence modeling levels to calculate risk. GeNIe, which was developed at the Decision Systems Laboratory at the University of Pittsburgh, is used to create the model [15].

4.0 Conclusions

This article briefly describes a framework based on Bayesian network for dynamic risk assessment for the oil/gas, chemical, and petrochemical industries. This research illustrates the capability of Bayesian networks to represent all component interactions in complex systems, and to update all prior system probabilities in the presence of new information. The advantages of modeling flexibility and efficient computational mechanisms using dynamic Bayesian networks are also discussed.

4. References

[1] Yang, X. and Mannan, M.S., "The development and application of dynamic operational risk assessment in oil/gas and chemical process industry", Reliability Engineering and System Safety 95 (2010) 806-815

[2] Khakzad, N., Khan, F. and Amyotte,P., "Safety analysis in process facilities: Comparison of fault tree and Bayesian network approaches", Reliability Engineering and System Safety 96(2011) 925-932

[3] Siu,N. "Risk assessment for dynamic systems: an overview", Reliability Engineering and System Safety 43 (1994) 43-73

[4] Devooght, J. and Smidts,C. "Probabilistic reactor dynamics-I: the theory of continuous event trees", Nuclear Science and Engineering 1992; 111:229-40

[5] Amendola, A. "Accident sequence dynamic simulation versus event trees", Reliability Engineering and System Safety 1988; 22:3-25

[6] Howard, R. "Dynamic probability systems", Vols. I and II, New York, 1971

[7] Sklet, S. "Comparison of some selected methods for accident investigation", Journal of Hazardous Materials 2004; 111:29-37

[8] Nivolianitou, Z.S., Leopoulos, V.N. and Konstantinidou, M., "Comparison of techniques for accident scenario analysis in hazardous systems", Journal of Loss Prevention in the Process Industries 2004; 17"467-75

[9] Pasman, H. and Rogers, W.; "BBN, a tool to make LOPA more effective, QRA more transparent and flexible, and therefore to make safety more definable", 14th Annual Sysposium, Mary Kay O'Connor Process Safety Center, College Station, Texas, October 2011

[10] Donohue, S.K. and Dugan, J.B.; "Modeling the "Good enough to release" decision usin V&V preference structure and Bayesian belief networks", Reliability and Maintainability Symposium, 2003 Annual, 2003

[11] McNaught, K.R. and Zagorecki, A. "Using Dynamic Bayesian networks for prognostic modeling to inform maintenance decision making", IEEM 2009, Hond Kong, January 2010

[12] Dean, T. and Kanazawa, K., "A model for reasoning about persistence and causation", Computational Intelligence 1989; 5:142-150

[13] Bobbio A, Portinale L, Minichino M and Ciancamerla E., "Improving the analysis of dependable systems by mapping fault trees into Bayesian networks", Reliability Engineering and System Safety 71(3): 249-260

[14] Bearfield G and Marsh W, "Generalising event trees using Bayesian networks with a case study of train derailment", Available online at http://www.agenarisk.com/resources/white_papers/Generalising_Event_Trees.pdf , last modified: 23 February 2010, accessed: 7 January 2012

[15] Decision Systems Laboratory, University of Pittsburgh, GeNIe (Graphical Network Interface and SMILE (Structural Modeling, Inference, and Learning Enginer), Version 2.0 software, 2010. Available online at http://genie.sis.pitt.edu

(137b) Comparing the Performance of Compact Heat Exchange Reactor to CSTR for Fischer Tropsch Synthesis

Hani Gadalla[1], Zhijun Jia[1], Steven J. Vallee[1], Burtron H. Davis[2]and Dennis Sparks[2], (1)Chart Energy & Chemicals, La Crosse, WI, (2)Center for Applied Energy Research, University of Kentucky, Lexington, KY

A commercial Fischer Tropsch catalyst was studied in two Compact Heat Exchange Reactors (CHER) with different channel sizes, and in a continuous stir tank reactor (CSTR). The CHERs were constructed by Chart Energy and Chemicals with one reactor having a 5mm channels and the other having a 6mm channels, while maintaining the same catalyst volumeThe catalyst was crushed and sieved retaining particles ranging in size from 75-125µm. Testing was performed at the University of Kentucky Center for Applied Energy Research in Lexington, KY. The CO conversion, selectivity and hydrocarbon productivity were measured for each reactor over a wide range of operating conditions, as a function of temperature, pressure, H2:CO feed ratio, and flow rate. Both reactor types showed excellent temperature control, but the performances varied at different operating conditions. The larger channel CHER had a higher conversion than the 5mm channel, potentially due to slightly higher reactor bed temperature. The two CHERs approached and even exceeded the productivity of the CSTR for liquid hydrocarbons per gram of catalyst. At higher flow rates the methane selectivity increases at a lower rate for either of the CHERs than for the CSTR.

(124a) Inspection War Stories: Part 1

Gregory A. Cantley, Marathon Petroleum Company LP, Ashland, KY

The recent inspection of multiple distillation columns within Marathon Petroleum Company demonstrates the need for a thorough inspection of tower internals prior to startup to avoid failure. A good understanding of the process and objectives of the tower are required to ensure that column internals will perform as expected.

(124b) Is the Hydraulic Gradient on Sieve and Valve Trays Negligible?

Henry Kister, Process Engineering, Fluor Corp., Aliso viejo, CA

Hydraulic gradient estimates are important for troubleshooting channeling and multipass maldistribution problem on trays, yet there is no satisfactory method in the open literature to evaluate it. A common assumption in sieve and valve tray calculations is that the hydraulic gradient is negligible. This is supported by the literature leading method for hydraulic gradient calculations. In contrast, a rule of thumb we published in the 8th edition of Perry's Handbook, based on quantitative gamma scan data analysis, suggests that the hydraulic gradient on these trays may be significant when the liquid load is high and the flow path is long.

This paper looks further into the prediction of hydraulic gradients by comparing to published FRI data.

(124c) Troubleshooting and Improving the Operation of a Main Fractionating Tower

Lowell Pless, Tracerco, Pasadena, TX and Shane LaCasse, CHS Refining, Laurel, MT

Following a refinery turnaround, the Main Fractionator in the Ultra-Low Sulfur Diesel unit was not able to operate at expected rates. The Main Fractionator was field tested using radioisotope technology and studied for not being able to operate as expected. Gamma scans identified two sources of flooding originating from a pair of draw downcomers. Neutron backscatter technique was used to further investigate and confirm downcomer hydraulics to identify the root-cause of flooding in the Fractionator. The downcomer draw design was studied and determined to be conducive for plugging and subsequent flooding. A simpler downcomer design was proposed and implemented renovating an existing complicated piece of hardware. In addition the experience resulted in revamped turnaround procedures to avoid a repeat episode in the future.

(124d) Troubleshooting a Benzene Splitter Capacity Limitation

Waldo E. De Villiers, Shell Global Solutions (US) Inc., Houston, TX, Joanne Chew, Shell Global Solutions, Malaysia and Daisuke Honbou, Seibu Oil Co., Ltd.

A Benzene/Toluene Splitter in the aromatics Solvent Extraction Unit of a Refinery was experiencing a capacity limitation. The maximum achievable throughput corresponded to tray loads that fell short of the expected flood point. Attempts to increase throughput would result in increased pressure drop in the top part of the column together with higher than desired toluene in the benzene side-draw. This presentation will review the test runs, diagnostic scans and other troubleshooting steps that were performed to better understand the column constraints. One of the key observations was an unexpected high liquid level on the benzene partial draw tray. Careful consideration of the test run and other diagnostic data led to a number of recommended tray modifications. These modifications were implemented and following start-up the column was able to achieve significantly higher throughput without increased toluene in the benzene. One of the interesting aspects of this case study was the interaction that was observed between the column and side-draw receiver due to the location of an equalizing line nozzle.

Unique Ultrasonic Measurement Solutions

Anand N. Vennavelli
Fractionation Research, Inc.
vennavelli@fri.org

Jon Cloy
Marketing, GE Sensing
Jon.Cloy@ge.com

Paper # 124e, 2012 AIChE Spring Meeting, Houston, TX. *

INTRODUCTION

Ultrasonic clamp-on technology could provide valuable troubleshooting information and is particularly attractive because of the non-invasive nature of the technology.

In September 2011, GE's ultrasonic clamp-on liquid flow meters were installed on the Reflux and Bottoms pipes at the Fractionation Research, Inc. (FRI) unit with the following goals:

- Evaluate non-invasive clamp-on ultrasonic flow technology.
- Test novel applications of ultrasonic technology such as correlating sound speed with liquid density or composition.

So far, FRI completed two tests with the GE liquid clamp-on meters — one each in September 2011 and November 2011. The C_6/C_7 system at 23.5 psia was used in both tests.

In this paper, the preliminary results of the two tests are discussed. The flow measurements from the GE clamp-on meters are compared to FRI's measurements of the liquid flow. Sound speed, also measured from the GE meter, is correlated with liquid density and composition.

DATA COLLECTION

In September 2011, two clamp-on ultrasonic flow meters were installed on the bottoms and the reflux pipes of the FRI unit. Figures 1 and 2 show the photographs of the clamp-on flow meters. A transmitter-indicator was set up from which the flow and the sound speed values could be read.

Flow and sound speed readings were synchronized with liquid sample collections such that all flow and sound speed measurements were taken at steady-state and corresponded to the composition and process data collected for FRI's tests. Mass flow rates measured by FRI's Coriolis flow meters were converted to volumetric flow rates for comparison.

---

*Unpublished. The AIChE shall not be responsible for statements or opinions contained in its publications.

RESULTS AND DISCUSSION

The range of the reflux and the bottoms flow rates for the two tests are shown in Table 1. The November 2011 test had a narrower range than the September 2011 test. The flow rate ranges for both tests were determined by the ongoing mass transfer program at FRI.

Table 1: Flow rate ranges for the 2011 GE Liquid Clamp-on Tests. The ranges were determined by the ongoing mass transfer program.

Clamp-on Test	Bottoms Flow GPM	Reflux Flow GPM
Sep 2011	175–400	9–60
Nov 2011	145–228	31–55

Flow Measurement Results

The GE ultrasonic liquid flow meter measures the flow rate by measuring the sound transit time through the liquid. The transit time is based on the difference in the time taken by sound to travel up-stream and down-stream. The volumetric flow rate can be read directly from the transmitter-indicator.

The volumetric flow rates measured by GE's flow meter and FRI's measurement for the reflux flow are compared in Figure 3 in a parity plot. The plot shows an excellent agreement between the liquid flow rates measured by the ultrasonic flow meter and FRI's Coriolis meter. The mean absolute relative error between the reflux flow ultrasonic and FRI measurements was 5% for the September test data 6% for the November test data.

Similarly, the volumetric flow rates measured by GE's flow meter and FRI's measurement for the bottoms liquid flow are compared in Figure 4 in a parity plot. The mean absolute relative error between the bottoms flow ultrasonic and FRI measurements was 13% for the September test data and 14% for the November test data. The scatter observed in the bottoms flow measurements was greater than that in the reflux flow measurements, but still within the initial expectation of 15%.

Sound Speed Measurement Results

The liquid ultrasonic flow meter indicates both the flow rate and the sound speed through the liquid. Throughout the test, when reflux and bottoms were sampled, the sound speed through the liquid was also recorded.

The volumetric flow rate is determined from the sound transit time characteristics. However, the sound speed value by itself could provide information about the process liquid. One of the R&D goals of this work is to explore a correlation between sound speed and liquid density and sound speed and composition. Sound speed through a liquid mixture is affected by both temperature and composition. After accounting for the effect of temperature changes, the variation in sound speed can be directly attributed to composition changes.

The reflux and the bottoms temperature ranges are shown in Table 2. For most pure liquids, sound speed is inversely proportional to temperature. A temperature correction to sound speed was implemented in this report assuming that the inverse relationship between sound speed and temperature is also valid for the

Table 2: Temperature ranges for the 2011 GE Liquid Clamp-on Tests. The ranges were determined by the ongoing mass transfer program.

Clamp-on Test	Bottoms Temperature °F	Reflux Temperature °F
Sep 2011	229–238	149–171
Nov 2011	229–234	135–154

C_6/C_7 binary mixture. The reflux sound speed was calculated using Equation (1) at a basis of 165°F and the bottoms sound speed using Equation (2) at a basis of 230°F. Note that the choice of the basis is arbitrary.

$$\text{Reflux sound speed (ft/s)} = \frac{\text{Recorded Sound Speed (ft/s)} \times \text{Reflux Temperature (°R)}}{625\ \text{(°R)}} \tag{1}$$

$$\text{Bottoms sound speed (ft/s)} = \frac{\text{Recorded Sound Speed (ft/s)} \times \text{Bottoms Temperature (°R)}}{690\ \text{(°R)}} \tag{2}$$

The positive correlation between the sound speed and liquid density of the reflux and the bottoms, first observed in the September test, was confirmed in the November test as shown in Figures 5 and 6.

Similarly, the positive correlation between sound speed and liquid C_6 composition of the reflux and the bottoms, first observed in the September test, was also confirmed as shown in Figures 7 and 8.

In summary, the sound speed measurements correlate well with liquid density and composition changes. The GE liquid clamp-on transmitter can potentially be used for the C_6/C_7 system to detect changes in liquid density or as a clamp-on on-stream analyzer!

CONCLUSIONS AND FUTURE WORK

The ultrasonic flow meters showed potential to provide liquid flow measurements with 5–15% accuracy using a non-invasive clamp-on technology.

According to GE, more accurate flow measurements (up to 2% and up to 0.25% in some cases) may be obtained by using flanged-in invasive wetted sensors. However, the goal of this work was to study the effectiveness of clamp-on technology.

Considering that the flow measurements have been obtained using a non-invasive clamp-on device, the technology is industrially very valuable, especially for troubleshooting. Furthermore, since the flow meter has no moving parts, little to no maintenance or calibration is required.

The results of the September and November tests together showed that the ultrasonic measurements are repeatable and consistent. The November test results support all conclusions from the September test. The ultrasonic technology showed potential for detecting changes in liquid density and composition and therefore potential for use as an on-stream analyzer for the C_6/C_7 system.

Additional experiments and testing of novel applications of ultrasonic technology are planned.

Figure 1: Photograph of the GE clamp-on ultrasonic flow meter on the reflux pipe.

Figure 2: Photograph of the GE clamp-on ultrasonic flow meter on the bottoms pipe.

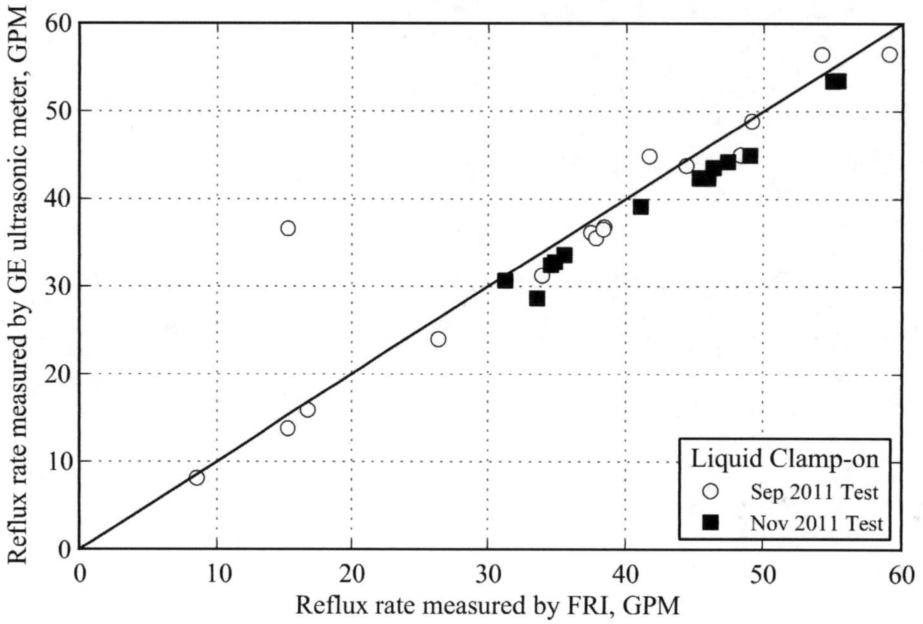

Figure 3: Parity plot of the reflux flow rate measurements

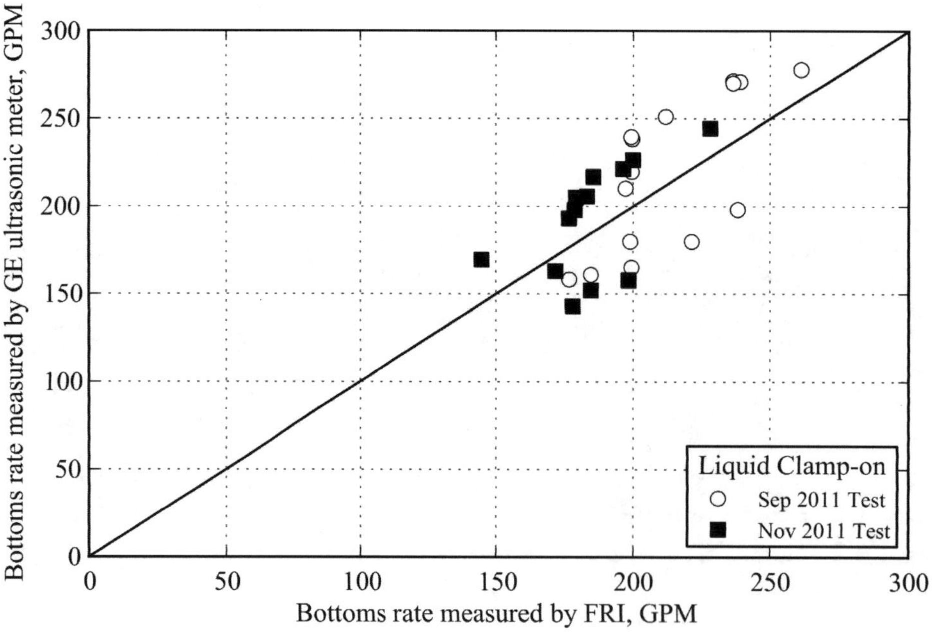

Figure 4: Parity plot of the bottoms flow rate measurements

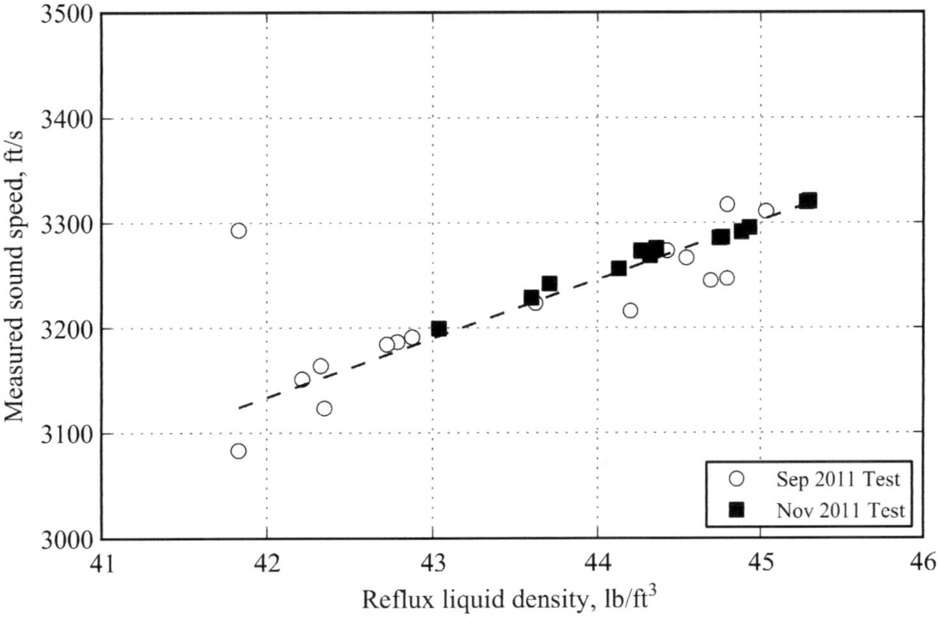

Figure 5: A positive linear correlation between the reflux sound speed and liquid density was observed. C_6/C_7 system at 23.5 psia. Reflux flow rates ranged from 9–60 gpm for the Septemer 2011 test and 31–55 gpm for the November 2011 test.

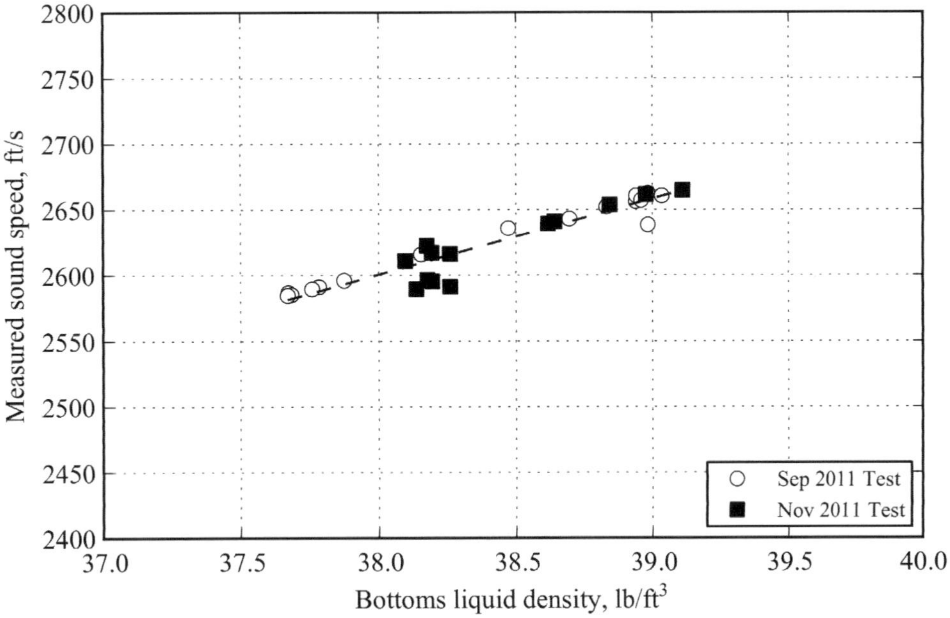

Figure 6: A positive linear correlation between the bottoms sound speed and liquid density was observed. C_6/C_7 system at 23.5 psia. Bottoms flow rates ranged from 175–400 gpm for the Septemer 2011 test and 145–228 gpm for the November 2011 test.

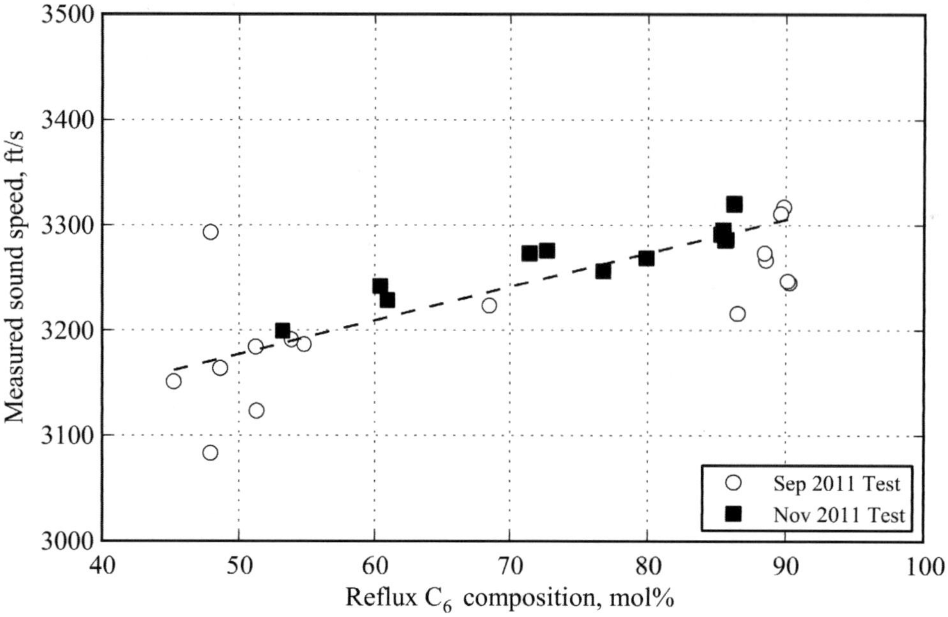

Figure 7: A positive linear correlation between the reflux sound speed and C_6 composition was observed. C_6/C_7 system at 23.5 psia. Reflux flow rates ranged from 9–60 gpm for the Septemer 2011 test and 31–55 gpm for the November 2011 test.

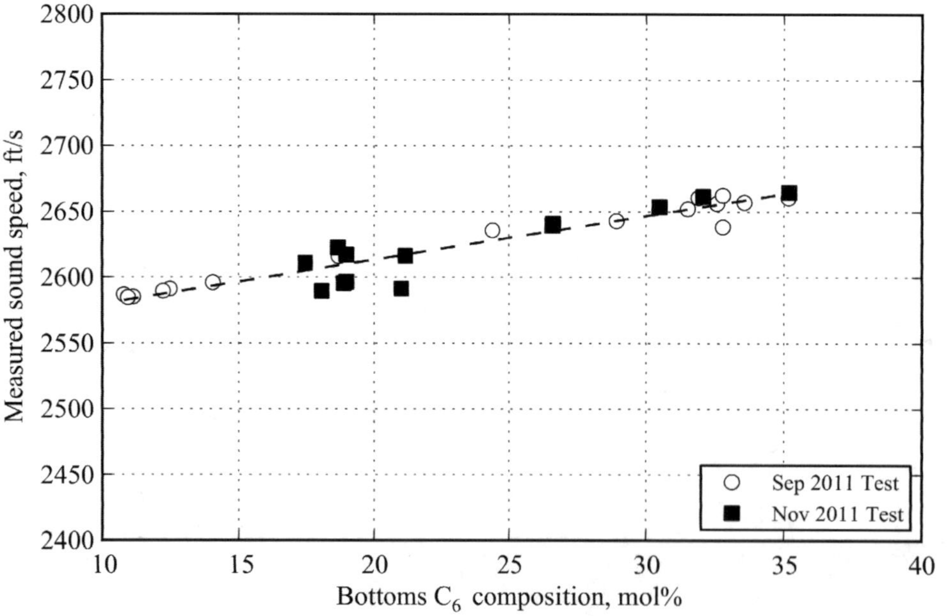

Figure 8: A positive linear correlation between the bottoms sound speed and C_6 composition was observed. C_6/C_7 system at 23.5 psia. Bottoms flow rates ranged from 175–400 gpm for the Septemer 2011 test and 145–228 gpm for the November 2011 test.

(124f) The Influence of Low Surface Tension on the Performence of Mist Eliminators

Wolfgang Heikamp, Rhodius GmbH, Weienburg, Germany, Michael Schultes, Raschig Ring Department, RASCHIG GMBH, Ludwigshafen, Germany and Attilio J. Praderio, LNG Technology Development, ConocoPhillips, Houston, TN

The influence of surface tension on the performance of mist separator

Wolfgang Heikamp, Rhodius GmbH

Attilio Praderio, Conoco Philipps

Prof. Dr. Michael Schultes, Raschig GmbH

In many chemical, refinery and petrochemical processes wire mesh mist separators are installed to remove fine droplets from gas streams. The design of these separators are mostly based on the following equation using a wire mesh characterizing K- factor,

$$Vc = K\sqrt{\frac{\Delta\rho}{\rho}} \quad \rho = \rho_L$$

Vc= design flow velocity and is a function of the L/V liquid/gas mass flow ratio, the density difference between liquid and gas $\Delta\rho$ and the liquid density ρ_L.

This equation (Souders - Brown) became an industrial standard to calculate a knitted wire mesh separators. There exist many proposals for the K factor in different applications.

Another design criteria was initially recommended by Brkholz, based on considerable test data, which can also be used for the estimation of the flooding velocity.

It has to be noticed that this equation does not take into account any surface tension effects and has been demonstrated only for air/water systems. Based on other gas/liquid separation processes it is known that low surface tensions (especially below 20 mN/m) reduce the flooding velocity.

In recent experimental studies with commercially available mist eliminators the influence of surface tension on the flooding velocity was investigated. A test rig was developed to study the flooding performance of mist eliminators with surface tensions from 68 mN/m down to 15 mN/m.

The results of these experimental investigations will be presented and compared to literature data. In addition an overview of future work will be given

(142a) Assessment of Minimum Energy and Operation Flexibility of Complex Dividing Wall Columns by Use of the Vmin-Diagram

Ivar J. Halvorsen, ICT, SINTEF, Trondheim, Norway, Sigurd Skogestad, Norwegian University of Science and Technology (NTNU), Trondheim, Norway, Igor Dejanovic, Chemical Engineering, University of Zagreb, Zagreb, Croatia and Zarko Olujic, Process & Energy Laboratory, Delft University of Technology, Delft, Netherlands

The dividing wall distillation columns (DWC) find increased use in industrial practice. The benefits are the combined savings in both energy and capital cost. Most of the industrial applications are three-product columns, but also 4-product DWCs have been constructed. A simulation study has shown that a four-product DWC with multiple partition walls for an aromatics feed can save up to 50% energy compared to conventional sequences. The internal arrangement is more complex and it is important to predict the necessary internal flow rates both for making proper sizing of the internals and also for control structure design. Thus, there is still a need to focus on development of simple-to-use methods for making assessment on how complex columns like the DWC will perform for any given application. This paper will demonstrate the usability of the Vmin-diagram method for quick assessment of column configurations for multicomponent separation.

The Vmin-diagram is obtained from feed data only, and gives a direct picture of the minimum vapor requirement for any specification of product splits. It is particularly well suited for fully thermally coupled distillation configurations like the generalized Petlyuk arrangements and dividing wall columns. The Vmin-diagram was first presented by Halvorsen and Skogestad (2003). The deduction was based on the simplifying assumptions of constant molar flows, constant pressure and constant relative volatility. It can also be constructed from simulations on a real mixture.

The minimum vapor flow rate in a DWC is given by the most difficult of the split between two adjacent product groups as if each split were to be carried out in a standard two-product column. For sharp split specifications, these are found as peaks in the Vmin-diagram. In addition, a base solution for the internal flow rates in each internal column section can be found directly.

In a DWC we fortunately have some tolerance in how to set the internal flows since there is normally more than a single optimal operating point. This flexibility arises when there are different minimum required flow rates in the upper and lower part of the arrangement. This tolerance can be explored in the Vmin-diagram and be used to specify operational margins on the internal flow rates. Knowledge of the operational tolerances for optimal operation is important in order to design and operate a column that obtains the theoretical energy savings in practice.

(142b) Experimental Evaluation of a Concentric Tray HIDiC Performance

Zarko Olujic, Sander Tromp, Joost Laven, Coen Dusseljee and Andrzej Stankiewicz, Process & Energy Laboratory, Delft University of Technology, Delft, Netherlands

An internally heat integrated distillation column (HIDiC) is essentially a heat pump assisted distillation column that in conjunction with diabatic operation, i.e. a uniform distribution of heat transfer duty along the integrated parts of a hot rectification and a cold stripping section, enables minimization of energy requirement in demanding separations. It is a proven, but regarding the hardware a highly demanding energy conservation concept. Indeed, uncertainties related to both design & construction and operation of a HIDiC are the main reason for reluctance on the side of potential users that diminishes prospects for practical implementation.

In present, technical feasibility study a concentric column containing an 0.3 m diameter rectification section placed inside an 0.8 diameter stripping section, with five fully thermally integrated sieve trays, has been used to simulate experimentally operation of a HIDiC. Two heat integrated sections have been operated as two individual columns in parallel at total reflux conditions, using appropriate mixtures of cyclohexane and n-heptane. The inner, rectification column (section) of a HIDiC was operated at higher pressures, while the outer, stripping section was operated at atmospheric pressure. The heat transfer from hot rectification column (section) to cold stripping column (section) occurred via the surface of the inner column wall and the heat panels placed above active (bubbling) area of the annular sieve trays in the stripping column. The photograph shows design of equipment.

The performance of the small diameter inner column (cartridge sieve trays) appeared to be more sensitive to the rate of internal heat transfer than the outer column, exhibiting a deteriorating trend in efficiency with increasing column vapor load. The outer column equipped with annular sieve trays exhibited a stable performance over the whole operating range and as expected the presence of heat transfer panels appeared to be beneficial to both hydraulics and mass transfer efficiency. However, the overall heat transfer coefficient exhibits a progressive increase with increasing vapor load, indicating that at lower loads insufficient wetting, i.e. vaporization occurs, which is a concern. This and other concentric tray HIDiC design and operation relevant considerations are discussed.

(142c) Kettle Reboiler Windows

T.J. Cai, Simon Chambers, A. Y. Ogundeji and A.N. Vennavelli, Fractionation Research, Inc., Stillwater, OK

Fractionation Research, Inc. (FRI) recently installed two windows on an industrial-size kettle reboiler. Using those windows, data were collected regarding pool depths, entrainment, flow patterns and nucleation sites. Several binary systems were employed. A wide range of steam pressures were employed. Video footage will be shown.

(142d) An Innovative Energy Saving Based on Self-Heat Recuperation for Distillation Process

Yasuki Kansha, Akira Kishimoto and Atsushi Tsutsumi, Institute of Industrial Science, The University of Tokyo, Tokyo, Japan

Although distillation columns have been widely used in separation processes based on vapor/liquid equilibria in petroleum refineries and chemical plants, the distillation process consumes a massive amount of energy required for the latent heat of the phase change, resulting in the emission of a large amount of CO_2.

To overcome this problem, an innovative energy saving based on self-heat recuperation for distillation processes is proposed in this research; the energy saving occurs via internal heat circulation of the process. It consists of a heat circulation module and a distillation module in which the heating and cooling loads are balanced by energy/exergy analysis. The self-heat recuperation technology proposed in our previous studies is adopted, and thus, the heat of condensation and the cooling heat are recuperated by compressors, and exchanged with the heat of vaporization and the heating heat by employing heat exchangers in both the heat circulation and distillation modules.

The self-heat recuperation technology helps to achieve a drastic energy saving for distillation of more than 80% in comparison with a benchmark distillation involving the use of an external heat source and a reboiler for heating.

APPLICATION OF RESPONSE SURFACE METHODOLOGY TO OPTIMIZE THE RETROFIT OF A SIDE COLUMN TO A DWC

Nguyen Van Duc Long and Moonyong Lee, Yeungnam University, Gyeongsan, South Korea

Introduction

Distillation plays an important role in the chemical process industries and consumes the largest amount of energy with an estimated 3% of the world's energy consumption [1]. Motivated by this large energy requirement for distillation, researchers have developed various column arrangements that can bring in savings in both energy and capital cost [2]. Any reduction of energy consumption will not only bring economical benefits but also environmental benefits in terms of reduced usage of fossil fuels and its associated emissions.

A side distillation column can be used to separate any nonazeotropic multicomponent mixture into three products [3]. Therefore, it can often substitute for a sequence of two columns. However, the purity of the side distillation is restricted by thermodynamics and by the nature of the distillation process. These columns are usually appropriate either as prefractionators (the side stream is fed to another column for further separation) or to generate recycle streams when there is no strict requirement of the recycle composition. They are also broadly used in the petroleum industry, where blends are produced rather than pure products.

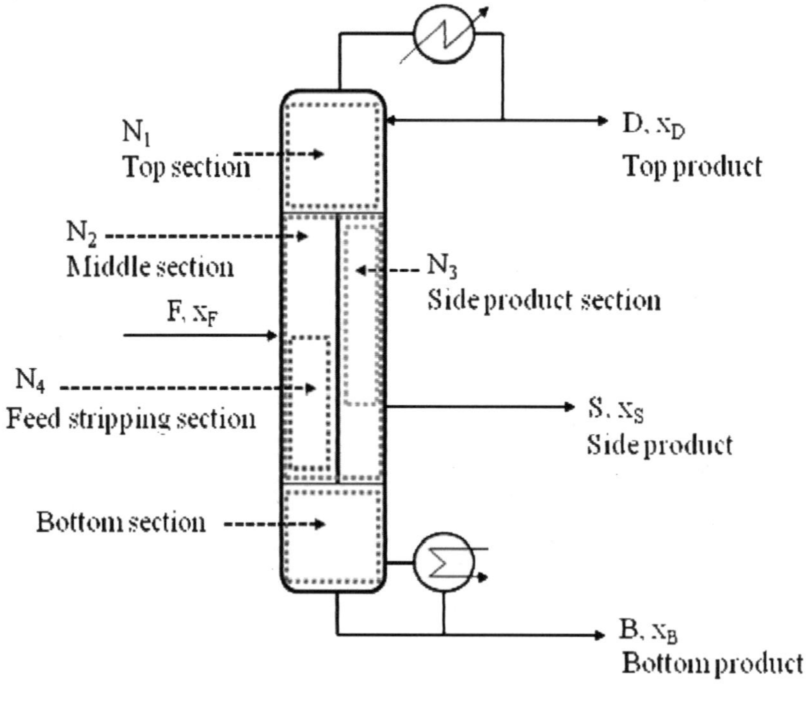

Figure 1. A dividing wall column.

There is a significant amount of literature analyzing the relative advantages of dividing wall column (DWC) options, schematically drawn in Fig. 1, offering a great chance at reduced energy consumption. These studies have shown that DWC systems are capable of achieving energy savings of up to 30% over conventional direct and indirect distillation sequences [4-8]. Besides, the investment and space saving can be gained because the number of reboilers, condensers and associated equipments such as pumps, their supports [2]. Therefore, DWC is attractive to many chemical and related industries in the current scenario of competition and environmental concerns, in order to reduce energy usage for distillation.

Retrofit a side distillation column to a DWC has potential for reducing energy. The successful revamp of a conventional tray Xylenes Distillation Column was accomplished by ExxonMobil [9]. The column takes a xylene sidestream from reformate to feed an aromatics plant to make higher value products. Changing the column configuration to a DWC resulted in improved product purity and significant energy savings. The project cycle is detailed including process design, hydraulic design, mechanical design, manufacture, testing and site work. An investment with payback period of less than 3 years is favorably considered by the management.

Recently, Premkumar & Rangaiah (2009) and Long (2010) studied the potential of retrofit conventional 2-column systems and side for separating ternary mixtures into three products to DWCs [2,7]. DWCs were optimized only one variable at a time, keeping constant the remaining ones. For each chosen variable value, the internal vapor and liquid flow to the prefractionator were varied to optimize energy consumption. The interactions between variables could not be identified and quantified by such a technique. Latterly, Long and Lee proposed the response surface methodology (RSM) and factorial design for DWC structure design and optimization, which is easy and efficient to implement using Hysys and Minitab [10-12]. These methods require a little computational effort while it allows control variables to be optimized simultaneously. Both of them develop the mathematical models that permit assessment of the relevance and statistical significance of variables being studied [13]. They facilitate the evaluation of interaction effects between variables. The aim of the present work is to show potentiality of retrofit side distillation column to a DWC. The response surface methodology was employed to minimize the reboiler duty.

Design and Optimization Methodology

Design

The key to a successful retrofit lies in maximizing utilization of the existing equipment, while simultaneously minimizing the new hardware so as to abbreviate capital costs [14]. The side column is now modified by adding a wall in the middle section while maintaining the same number of trays. Initial DWC structure design involved applying a shortcut design procedure based on a conventional column configuration [7,12]. In this conventional configuration, the first column corresponds to the prefractionator section in the DWC. The rectifying section of the second column and the stripping section of the third represent the top and bottom sections of the DWC, respectively. The stripping section of the second column and the rectifying section of the third are equivalent to the dividing wall section of the DWC. The bottom stream from the second column and the top stream from the third column correspond to the side stream of the DWC. Consequently, the DWC structure can be divided into four sections: the prefractionator section for feed mixture, the top and bottom sections above and below the dividing wall section, and the dividing wall section.

Response surface methodology

Response surface methodology (RSM) is also employed to optimize the retrofit the side column to dividing wall column. After determining the preliminary ranges of the variables through single-factor testing, a Box-Behnken design was employed under the response surface methodology to analyze how the variables interacted and optimize the system in terms of reboiler consumption and cost saving. Simulation run data were fitted to a second-order polynomial model and regression coefficients were obtained. The generalized second-order polynomial model used in the response surface analysis is as follows:

$$Y = \beta_0 + \sum_{i=1}^{k} \beta_i X_i + \sum_{i=1}^{k} \beta_{ii} X_i^2 + \sum_{i<j}\sum \beta_{ij} X_i X_j + \varepsilon \quad (1)$$

where Y is the predicted response, X_i are the uncoded or coded values of the variables, β_0 is a constant, β_i, β_{ii} and β_{ij} are the coefficients of the linear, quadratic and interactive terms, respectively, and ε is the error term.

Case Study

Existing Side Distillation Column

Figure 2 illustrates the existing distillation column and current operating conditions. The existing side column is valve-trayed type of 3.8 m diameters, with 65 trays [15]. The feed composition, temperature, and pressure are described in Table 1. The simulation work was performed using simulator Aspen HYSYS V7.1. The NRTL-HOC property method that uses the Hayden-O'Connell equation of state as the vapor phase model and NRTL for the liquid phase was used for the prediction of the vapor-liquid equilibrium (VLE) of these simulations. Dimerization affects VLE, vapor phase properties, such as enthalpy and density, and liquid phase properties, such as enthalpy. The Hayden-O'Connell equation reliably predicts the solvation of polar compounds and dimerization in the vapor phase that occurs with mixtures containing carboxylic acids [16]. From the base case simulation model, it shows that the energy consumption of this column is 22.50 MW.

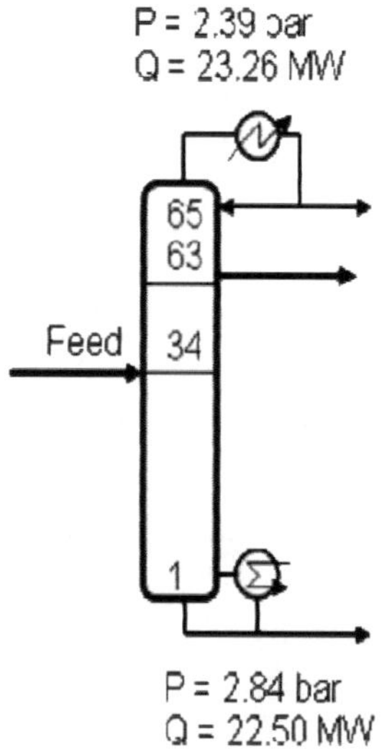

Figure 2. Simplified flow sheet illustrating the existing side distillation column.

Column hydraulics, as well as flooding, has to be considered. When the downcomer is small or the downcomer backup is high, downcomer flooding occurs, causing liquid accumulation in the tray above, while the rising of the vapor velocity causes entrainment. To determine the max flooding of a particular column, the rating mode is simulated with column internal specifications such as type of trays, column diameter, tray spacing, and number of passes. The side column was designed with a load which is at near 85% of the load at the flooding point to prevent flooding in the columns.

Table 1. Feed mixture condition and product requirement

Feed Conditions	
Component	Mass Flow (kg/h)
Formic Acid (FA)	36.29
Water	15.88
Acetic Acid (AA)	51848.41
Propionic Acid (PA)	313.43
Temperature (°C)	173.8
Pressure (bar)	4.46
Product Requirement	
Formic Acid (FA)	<1000 ppm
Water	<1000 ppm
Propionic Acid (PA)	<300 ppm

Dividing wall column

Table 2. Factors' coded levels for RSM

Factor	Levels		
	-1	0	1
top section (N1)	1	5	9
middle section (N2)	36	43	50
Side product location (N3)	1	3	5
Feed stripping location (N4)	1	5	9

In this paper, the retrofit project purpose is to reduce energy requirement, with the following assumptions:

a. The purity and recovery of AA is kept constant.
b. Specification of product: - FA and water concentration must be less than 1000 ppm
   - PA concentration is less than 300 ppm
c. Steam @ 150 psig (11.36 bar): 880 USD/KW.year [17]
d. Remained year: 8 years
e. Column material: stainless 316
f. New trays are installed in the dividing wall section

Table 2 shows the factors and levels used in this case study. 27 simulations were run to optimize 4 parameters of the DWC structure. For each run, internal vapor (F_V) and liquid (F_L) flows to the prefractionator were varied to minimize the reboiler duty. The minimum reboiler duty was found at coded levels of the number of trays in the top, middle, side product, and feed stripping section of -1.000, -0.3737, -1.000 and -1.000, respectively (Fig. 3).

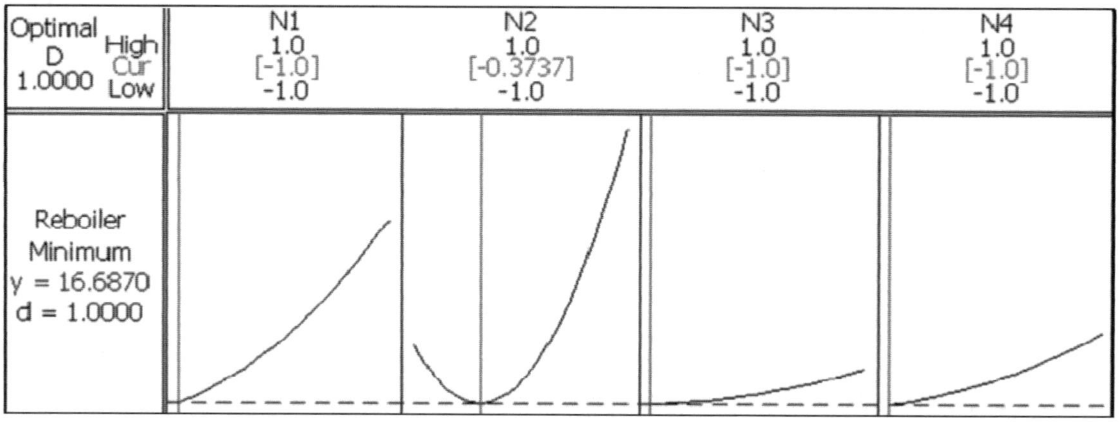

Figure 3. Optimization plot by the RSM.

Under these conditions, the minimum reboiler duty was predicted as 16.69 MW. The natural values of the valuables can be derived from the coded levels. F_V and F_L then were optimized to maximize the total annual cost saving, while maintaining sufficient product purity. Figure 4 shows a simplified flow sheet illustrating the resulting DWC system. The result from rigorous simulation also revealed that the DWC can save up to 25.16% of energy consumption, which showed good agreement with the predicted result by the RSM. The predicted maximum costs saving by the DWC is 3.89 million USD after 8 years. Simple payback period can be calculated as additional capital costs divided by saving per year. In this case the value is 2.2 months due to high energy cost. The chemical/petrochemical plants usually have an annual maintenance period of 10 – 14 days, during which major repair and replacement are carried out. The retrofitting discussed above has been assumed to take place within this maintenance period, thus production loss during the retrofitting time is not considered [2]. Note that the temperatures on the bottom section of the side column and DWC were almost same (165°C), medium pressure steam are used in both columns. Max flooding reduces from 87% to 67%, i.e. retrofit of side distillation column gives some chances for increasing the plant production capacity.

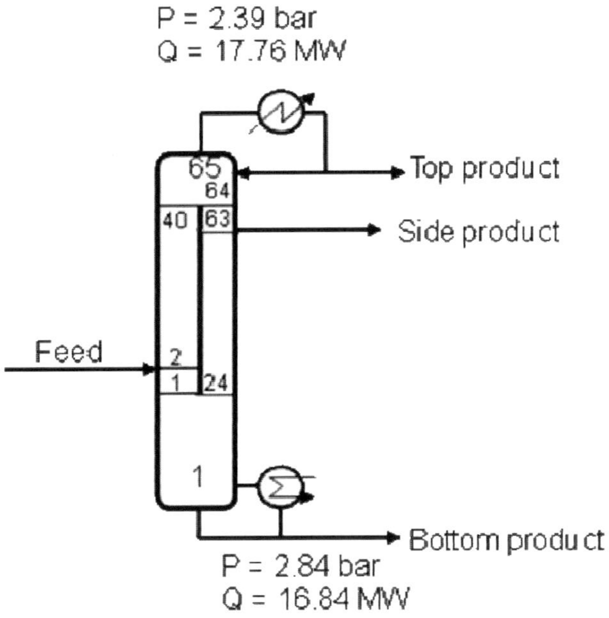

Figure 4. Simplified flow sheet of the resulting DWC system using RSM.

Conclusions

This paper has focused upon the retrofit the side column to a DWC, which maximizes utilization of the existing equipments and operate the plant more efficiently, possibly with minor modifications and small investment. The shortcut method, utilizing the three conventional column configuration structurally equivalent to the DWC, was used to determine the proper initial structure for each DWC in a simple manner. The response surface methodology was employed for retrofit design and optimization. The optimal structure design of the DWC has been considered in terms of the internal recycle flow distribution around the dividing wall, the number of trays in the top, middle, side product and feed stripping sections. One case study was analyzed and the results showed that RSM is easy and efficient to employ for retrofit design and optimize. The retrofitting DWC structure was efficiently optimized with a little computational effort. The predicted results by the RSM showed a good agreement with the actual ones by rigorous simulation. The resulting DWC leads to significantly reduced energy costs over side distillation column.

Acknowledgments

This work also was supported by the Human Resources Development Program of Korea Institute of Energy Technology Evaluation and Planing (KETEP) grant (No. 20104010100580) funded by the Korean Ministry of Knowledge Economy.

References

1. G. Hewitt, J. Quarini, M. Morrell, More efficiency distillation, *Chem. Eng.* (1999).
2. R. Premkumar, G.P. Rangaiah, Retrofitting conventional column systems to dividing wall columns, *Chem. Eng. Res. Des.* 87 (2009) 47-60.
3. K. N. Glinos, M. F. Malone, Design of sidestream distillation columns, *Ind. Eng. Chem. Process Des. Dev.,* 24 (1985) 822-828.
4. Victoria E. Tamayo-Galván, Juan Gabriel Segovia-Hernández, Salvador Hernández, Julián Cabrera-Ruiz, J. Rafael Alcántara-Ávila, Controllability analysis of alternate schemes to complex column arrangements with thermal coupling for the separation of ternary mixtures, *Com. Chem. Eng.*, 32, 3057-3066, 2008.
5. Hern´andez, S., & Jim´enez, A. (1999a). Design of energy-efficient petlyuk systems. *Computers and Chemical Engineering*, 23(8), 1005.
6. Sotudeh, N., & Shahraki, B. H. (2008). Extension of a method for the design of divided wall columns. *Chemical Engineering Technology*, 31, 83.
7. N. V. D. Long, Lee, S. H. & Lee, M. Y. (2010). Design and optimization of a dividing wall column for debottlenecking of the acetic acid purification. *Chemical Engineering and Processing*, 49(8), 825-835.
8. C. Triantafyllou, R. Smith, The design and optimization of fully thermally coupled distillation, *Trans. Inst. Chem. Eng.* 70 (Part A) (1992) 118-132.
9. B. Slade, B. Stober, D. Simpson, Dividing wall column revamp optimizes mixed Xylenes production, *IChemE Symp.* Series 2006, 152.
10. N. V. D. Long and M. Y. Lee, Dividing wall column structure design using response surface methodology, *Computers & Chemical Engineering*, 37 (2012) 119-124.
11. N. V. D. Long and M. Y. Lee, Design and optimization of a dividing wall column by factorial design, *Korean Journal of Chemical Engineering*, In Press (2012).
12. N. V. D. Long and M. Y. Lee, Improvement of the Deethanizing and Depropanizing Fractionation Steps in NGL Recovery Process Using Dividing Wall Column, *Japan Journal of Chemical Engineering*, In Press (2012).
13. César Ricardo Teixeira Tarley, Gustavo Silveira, Walter Nei Lopes dos Santos, Geraldo Domingues Matos, Erik Galvão Paranhos da Silva, Marcos Almeida Bezerra, Manuel Miró, Sérgio Luis Costa Ferreira, Chemometric tools in electroanalytical chemistry: Methods for optimization based on factorial design and response surface methodology, *Microchemical Journal* 92 (2009) 58–67.
14. K.A. Amminudin, R. Smith, Design and optimization of fully thermally coupled distillation columns. Part 2: Application of dividing wall columns in retrofit, *Trans. IChemE.* 79 (2001) 716-724.
15. PEP Report, Acetic Acid by Low Pressure Carbonylation of Methanol, November, 1994.
16. Aspen Technology, Aspen Physical Property System, version number V7.1, January 2009.
17. Biegler, L.T., Grossmann, I.E. & Westerberg, A.W. (1997). Systematic methods of chemical process design. Prentice Hall Inc., New Jersey, 110-141.

A STUDY ON SEPARATION PROCESS IN FLNG

Nguyen Van Duc Long and Moonyong Lee, Yeungnam University, Gyeongsan, South Korea

Introduction

The primary use of natural gas (NG) is as a fuel; it can also be a source of hydrocarbons for petrochemical feed stocks and a major source of industrially important elemental sulfur [1]. Its clean burning and ability to meet stringent environmental requirements have raised the demand for natural gas [2], which is supporting the rapid growth of liquefied natural gas (LNG) production capacity. Much of the world's gas reserves are in offshore fields [3], though onshore LNG processing is generally favored [4-8]. Recently, offshore FLNG (floating LNG) and LNG-FPSO (floating, production, shipping and offloading) services are beginning to be explored instead of land-based LNG plants [9]. FLNG plants must be safe, compact and energy efficient.

Several analyses of dividing wall columns (DWCs, Figure 1) for ternary separations have shown that they can achieve energy and space savings of up to 30% over conventional direct and indirect distillations [10-12]. DWCs allow reversible splits with no part of the separation performed twice, the main reason for their superior energy efficiency [13]. However, their design is more complex than conventional arrangements because of the greater number of degrees of freedom [14] that interact with each other and need to be optimized simultaneously.

Optimization studies rely on statistical approaches, with response surface methodology (RSM) being routinely used in several biotechnological and industrial multivariable processes [15-16]. In RSM, the Box-Behnken design only has three levels (low, medium, and high, coded as -1, 0, +1) and requires a small number of experiments or simulation runs. It is more efficient and easier to arrange and interpret than other methods.

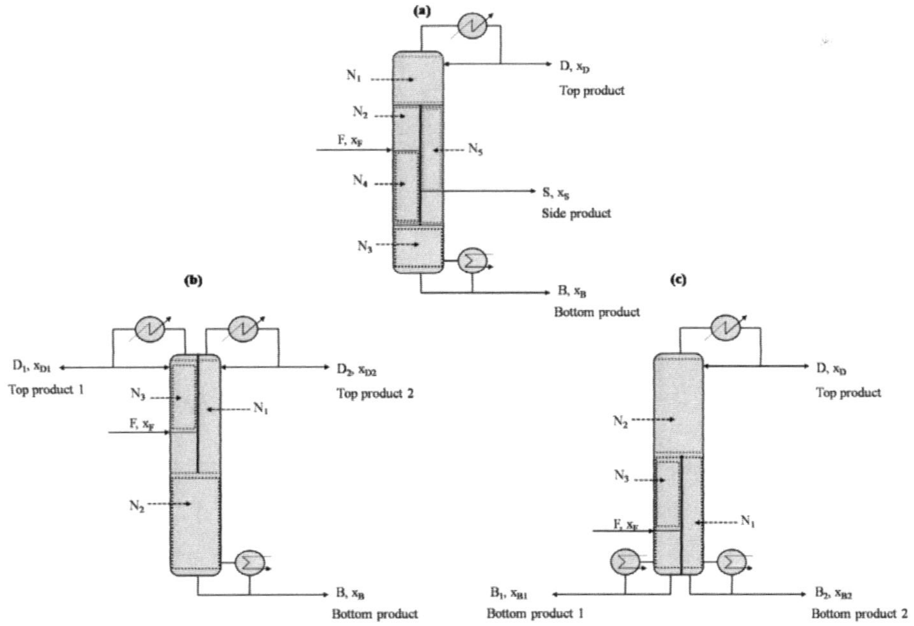

Figure 1. Dividing wall columns (DWCs): (a) conventional DWC; (b) top DWC; (c) bottom DWC.

This work proposes a new, simple, compact and power-efficient NGL recovery sequence for offshore FLNG. In the proposed NGL recovery process, feed splitting and a top dividing wall column were employed to maximize energy efficiency and plant compactness.

Analysis of Onshore LNG and Offshore FLNG Plants

A traditional onshore plant for offshore gas reserves usually includes an offshore platform for dehydration and compression, large pipelines to shore, an onshore liquefaction plant, and a harbor to accommodate purpose-built LNG carriers [17]. A floating LNG (FLNG) plant eliminates the platform, pipelines, and harbor, greatly reducing infrastructure requirements. The savings can be significant depending on the length of the pipeline and the cost of the harbor. Even for onshore gas fields, FLNG may also be less expensive than onshore facilities if extensive site preparation is required and there is no local infrastructure or natural harbor. It can also lessen a facility's environmental impact, particularly if the potential onshore plant would be located near populated areas. A FLNG plant can be moved to a new location after the depletion of the original gas field [18].

However, a FLNG facility's production capacity is limited by the deck space required for equipment, constraining throughput and economics, especially for projects on larger gas fields [19]. FLNG projects are more difficult to expand, as expansion is only achievable through the installation of additional 'stand alone' facilities. The expansion of a onshore plant would require less capital and materials, with only incremental additional environmental impact as additional common-use infrastructure such as power generation or export jetties may not be needed. A FLNG plant would have higher operating and maintenance costs than an onshore plant, with very limited operational synergies available for multiple FLNG facilities compared with an onshore plant operating multiple gas trains. FLNG plants are therefore most suited for developing small, remote gas fields that are difficult to develop commercially by conventional methods (often referred to as stranded gas assets).

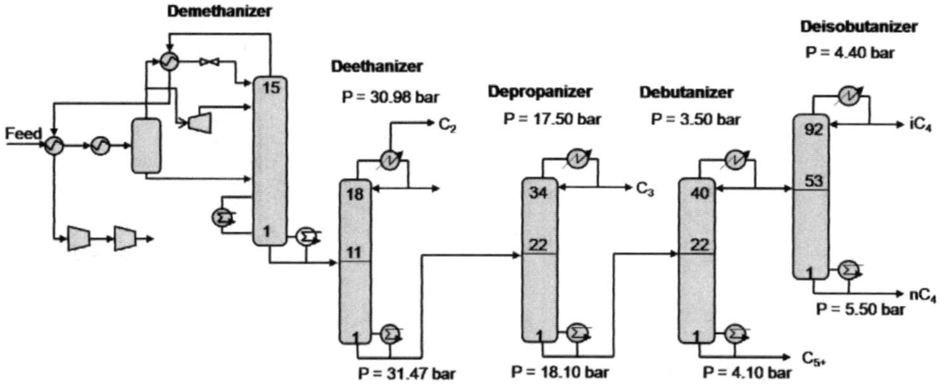

Figure 2. The separation train of typical onshore NGL processing.

Onshore liquefaction plants employ complex, high-efficiency liquefaction processes [20] that use mixed or cascade refrigerants with separate pre-cooling. However, offshore applications require simpler systems due to the challenges of moving processes offshore [18]. An offshore plant must be able to withstand waves, most notably the distillation and separation columns required for condensate stabilization, LPG extraction and acid gas removal in feed gas pretreatment. FLNG plants require higher levels of safety due to their compactness; interlocks, emergency shutdown, fire and gas detection, and fire fighting systems should be carefully considered from the conceptual stage [20].

Heat exchangers, cold-boxes and columns should be designed to withstand motion, making packed type column designs suitable [20]. This paper considers the potential process and the improvements that could be realized by separation. During onshore processing, liquid hydrocarbons recovered from NGL are typically separated into relatively pure ethane (C_2), propane (C_3), isobutane (iC_4), normal butane (nC_4), and gasoline (C_{5+}) products. This is conventionally done by the sequential distillation of C_2, C_3 and C_4 from gasoline, followed by further distillation of iC_4 from nC_4 (Figure 2) [4-8]. The similar boiling points of iso- and normal butane make their separation require many trays with a high reflux ratio. Due to restrictions of column height, a deisobutanizer would likely be excluded from an FLNG plant. In addition, the number of products and product tanks should also be minimized because of the limited space and safety concerns.

Proposed Base Sequence

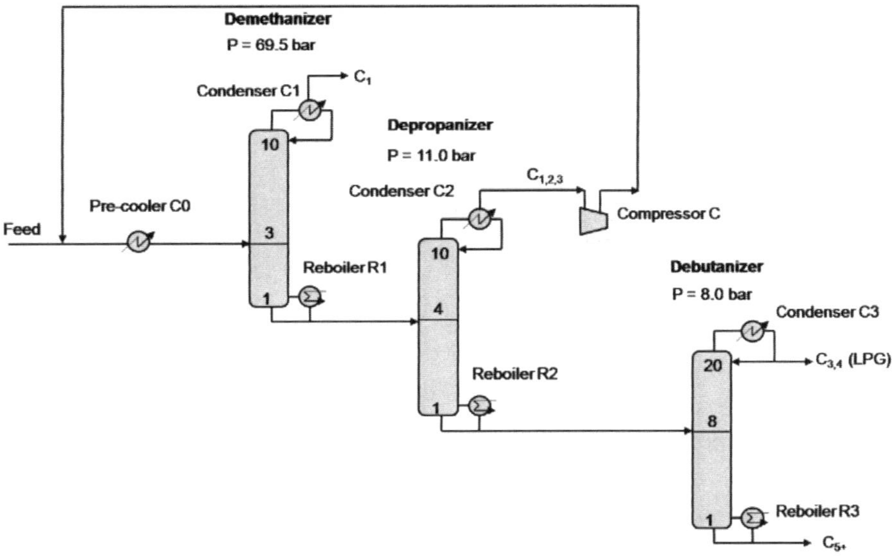

Figure 3. The separation train of three conventional columns.

A new NGL recovery process is proposed for FLNG facilities with the following constraints and assumptions:

a. The maximum allowable number of trays of each column is limited to 21 to limit the motion of wind and waves and to avoid excessive bending moments.
b. All columns are designed as packed type columns for greater stability against motion.
c. The operating velocity of all columns is near 80% of the flooding velocity.
d. The product specifications are as follows:
 - C_1: 89% (top of demethanizer)
 - LPG: 98% (top of debutanizer)
 - C_{5+}: 99% (bottom of debutanizer)

Based on the feed conditions and the product specifications, simulations were performed to quantify energy consumption. The demethanizer, with 10 theoretical trays, was designed for use at 69.5 bar (Figure 3). The sharp separation of methane from C_{2+} increases energy consumption. Therefore, a depropanizer column is used to recycle light components in the bottom stream of the demethanizer. The 20-tray debutanizer column was designed for use at 8.0 bar, as commercial LPG can be condensed with cooling water at this pressure.

The Peng-Robinson equation of state supporting the widest range of operating conditions and the greatest variety of systems was used to predict the simulations' vapor-liquid equilibria [21]. The Lee-Kesler method of enthalpy calculation was employed.

Improvement of Proposed Base Sequence

Feed Split

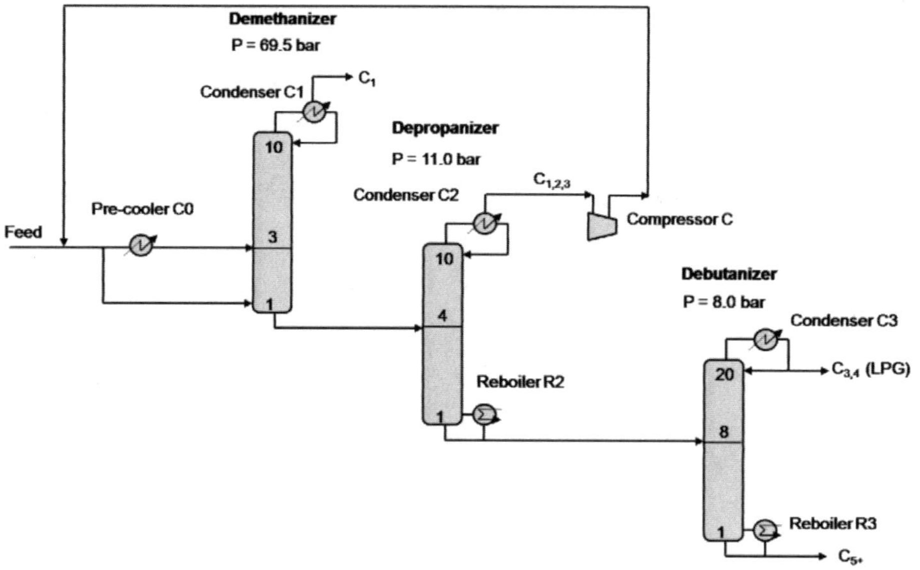

Figure 4. The separation train of three conventional columns with feed splitting.

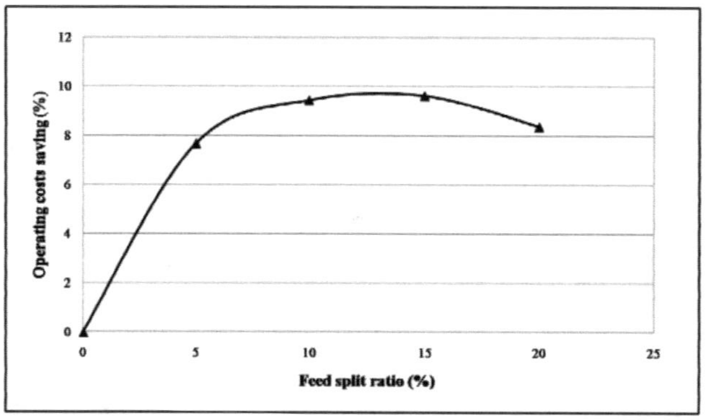

Figure 5. Effect of feed split ratio on the operating cost of the demethanizer and depropanizer.

Instead of pre-cooling the entire feed stream, it can first be divided into two streams: one is cooled before entering the column like a feed, and one that is fed into the column's bottom (Figure 4). This sequence can eliminate one reboiler in the demethanizer and save energy in pre-cooler C0. However, energy consumption is increased in condenser C1 in the demethanizer and the reboiler in the depropanizer. Note that the refrigeration used in precooler C0 and condenser C1 are at low temperatures. Therefore, adjustment of the feed split ratio is important in the optimization (Figure 5). A feed split ratio of 15% is optimal, reducing reboiler duty and operating costs by 47.43% and 9.64%, respectively.

Top Dividing Wall Column (TDWC)

In this case, two different cooling sources, water and refrigeration, are employed. Due to the high cost of refrigeration, integrating two columns into one dividing wall column with only one condenser is not efficient. Therefore, a top dividing wall column is considered here to integrate the depropanizer and the debutanizer for improved energy efficiency.

Optimization Methodology

The main design variables of internal vapor flow to the prefractionator (F_V), and the numbers of trays in the top (N1), botttom (N2), and feed rectifying (N3) sections were optimized. The objectives of the optimization were the minimization of the total number of trays (N) and the reboiler duty (Q). The optimization was constrained by the product purities and recoveries.

$$\min(Q, N) = f(N1, N2, N3, F_V) \quad (1)$$

Multiple response optimization can be undertaken by forming a constrained optimization problem with one of the responses treated as the objective of a constrained optimization problem and other responses treated as the constraints where the constraints' boundaries are determined by the process design engineers. Here, reboiler duty is considered the objective. The numbers of trays are the constraints and depend on the design engineers. Note that the number of trays in each column must be less than 21 trays.

A Box-Behnken design was employed under the response surface methodology to analyze how the variables interacted and to optimize the system in terms of reboiler duty and the number of trays. After determining the variables' preliminary ranges through single-factor testing, simulation run data were fitted to a second-order polynomial model and regression coefficients were obtained. The generalized second-order polynomial model used in the response surface analysis is as follows:

$$Y = \beta_0 + \sum_{i=1}^{k} \beta_i X_i + \sum_{i=1}^{k} \beta_{ii} X_i^2 + \sum_{i<j}\sum \beta_{ij} X_i X_j + \varepsilon \quad (2)$$

where Y is the predicted response, X_i are the uncoded or coded values of the variables, β_0 is a constant, β_i, β_{ii} and β_{ij} are the coefficients of the linear, quadratic and interactive terms, respectively, and ε is the error term. MINITAB software was used for response surfaces fitting.

TDWC Configuration

After initially setting the TDWC structure using a shortcut method [12,14], it was then optimized using response surface methodology. The DWC parameters were optimized over 15 simulation runs. For each run, the internal vapor flows to the prefractionator were varied to meet the required product purity and recovery. The optimum reboiler duty saving corresponds to the number of trays, increasing rapidly as the number of trays increases from 19 to 22. Therefore the maximum allowable 20 trays were employed in the design of the TDWC.

The greatest reboiler duty saving (11.89%) was predicted to have coded levels of -0.8192, 0.8192, and -1 corresponding to the numbers of trays in the top, bottom, and feed rectifying sections, respectively. The variables' natural values can be derived from the coded levels. The optimized TDWC system (Figure 6) had a simulated saving of 11.79%, in good agreement with the predicted value.

This column can be used to integrate two columns, whose condensers are cooled by different coolants; particularly the the depropanizer requires refrigeration, while the debutanizer can condense the top vapor stream under water cooling. The simulation results show that the TDWC can save up to

35.09% of refrigeration costs compared with a conventional column sequence. The duties of condenser (C3) and the reboiler can be reduced by 21.42% and 11.79%, respectively. Using a 1.5 m diameter TDWC could reduce energy consumption by 23.62% in terms of TAC compared with a conventional column sequence. Therefore, this kind of column can reduce energy consumption (refrigeration energy and reboiler and condenser duty) and investment costs (replacing two columns with one and eliminating a reboiler).

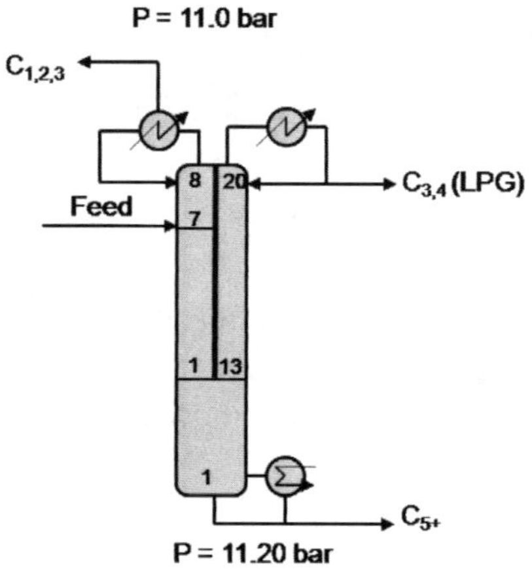

Figure 6. The TDWC system.

Conclusions

A new, energy-efficient and compact NGL recovery is proposed for offshore FLNG plants. Feed splitting could eliminate the demethanizer reboiler and reduce precooling energy. Operating costs were minimized at a feed split ratio of 15%. A TDWC allowed more energy-efficient depropanization and debutanization. A shortcut method determined a suitable initial structure of the TDWC, which was then optimized by response surface methodology. The proposed method could be simply and efficiently implemented in HYSYS and Minitab. Simulated results agreed well with predicted values. This kind of column could reduce energy consumption and investment costs. The reduced numbers of columns and reboilers could also make a plant more compact, important for offshore facilities.

Acknowledgments

This research was supported by a grant from the Gas Plant R&D Center funded by the Ministry of Land, Transportation and Maritime Affairs (MLTM) of the Korean government.

References

1. Kidnay, A. J.; Parrish W. R. *Fundamentals of Natural Gas Processing*; Taylor and Francis: Baca Raton, 2006.
2. Elliot, D.; Qualls, W. R.; Huang S.; (Roger) Chen J. J. Benefit of Integrating NGL Extraction and LNG Liquefaction Technology. *AIChE Spring National Meeting, 5th topical conference on Natural Gas Utilization (TI) Session 16c-Gas,* **2005**.
3. Sheffield, J. A. Offshore LNG Production – How to Make it Happen, Business Briefing: LNG

Review 2005.
4. Long N. V. D.; Lee M. Y. Improvement of Natural Gas Liquid Recovery Energy Efficiency through Thermally Coupled Distillation Arrangements, *Asia Pac. J. Chem. Eng.*, In Press, **2011**.
5. Manley, D. B. Deethanizer/Depropanizer Sequences with Thermal and Thermo-Mechanical Coupling and Component Distribution. U.S. patent 5,673,571, **1997**.
6. Mak, J. Configurations and Methods for Improved NGL Recovery. U. S. Patent 7,051,552, **2006**.
7. Long, N. V. D.; Lee, M. Y. Reduce Costs and Energy Consumption of Deethanizing and Depropanizing Fractionation Steps in NGL Recovery Process. *Computer Aided Chemical Engineering*, **2011**, *29*, 1658-1662.
8. Manley, D. B. Multiple Effect and Distributive Separation of Isobutane and Normal Butane. U.S. patent 8,806,339, **1998**.
9. Lee, S.; Cha, K.; Park, C.; Lee, C.; Cho, Y.; Yang, Y. The Study on Natural Gas Liquefaction Cycle Development. *ISOPE PACOMS,* Busan, Korea, **2010**.
10. Tamayo-Galván, V. E.; Segovia-Hernández, J. G.; Hernández, S.; Cabrera-Ruiz, J.; Alcántara-Ávila, J. R. Controllability Analysis of Alternate Schemes to Complex Column Arrangements with Thermal Coupling for the Separation of Ternary Mixtures. *Com. Chem. Eng.* **2008**, *32*, 3057-3066.
11. Hern´andez, S.; Jim´enez, A. Design of Energy-Efficient Petlyuk Systems. *Com. Chem. Eng.* **1999**, *23*, 1005-1010.
12. Long, N. V. D.; Lee, S. H.; Lee, M. Y. Design and Optimization of a Dividing Wall Column for Debottlenecking of the Acetic Acid Purification Process. *Chem. Eng. Process.* **2010**, *49*, 825-835.
13. Poth, N.; Brusis, D.; Stichlmair, J. Minimaler Energiebedarf von Trennwandkolonnen. *Chem. Ing. Tech.* **2004**, *76*, 1811–1814.
14. Amminudin, K. A.; Smith, R.; Thong, D. Y. C.; Towler, G. P. Design and Optimization of Fully Thermally Coupled Distillation Columns: Part 1: Preliminary Design and Optimization Methodology. *Trans. IChemE.* **2001**, *79 (Part A)*, 701-715.
15. Box, G. E. P.; Behnken, D. W. Some New Three Level Designs for the Study of Quantitative Variables. *J. Technometrics*, **1960**, *2*, 455-475.
16. Long, N. V. D.; Lee, M. Y. Dividing Wall Column Structure Design using Response Surface Methodology. *Comp. Chem. Eng.* **2012**, *37*, 119-124.
17. Naklie, M. Mobil's Floating LNG Plant. *International Conference & Exhibition on Liquefied Natural Gas 12th*, Perth, Australia, 4-7 May, **1998**.
18. Kerbers, I.; Hartnell, G. A Breakthrough for Floating LNG. *24th World Gas Conference*, Argentina, 5-9 October, **2009**.
19. Shell Development (Australia) Proprietary Limited (Shell), Prelude Floating LNG Project, October, **2009**.
20. Lee, Y. M.; Cho, T. I; Lee, J. H.; Kwon, O. Y. Trends and Technologies in LNG Carriers and Offshore LNG Facilities. *Offshore Technology Conference*, Houston, Texas, U.S.A, 5-8 May **2008**.
21. Aspen Technology. Aspen HYSYS Thermodynamics COM Interface, version number V7.1, January 2009.

Simulation and Optimization of Rectisol Using Revised Thermodynamic Methods

Lili Liu[1], *Fang Cai*[1], ***Jinsheng Sun***[1], *Wenchao Han*[2]
[1]*School of Chemical Engineering and Technology, Tianjin University, China*
[2]*China Special Equipment Inspection and Research Center (CSEI), Beijing, China*

1. Introduction

Rectisol is an economical, eco-friendly and well-proven process for the removal of acid components from raw gases, such as syngas and natural gas [1]. It uses hypothermal methanol as a solvent because it has a high vapor pressure and a strong absorptive capacity for acid gases in the normal operation range of -30 to -100℉(-34.4℃to -73.3℃) [2]. Rectisol technology provides an excellent option for the co-removal of many contaminants to downstream catalysts in one integrated unit. Contaminants including H_2S, COS, CO_2, and organic sulphides, were reduced to trace levels (e.g. H_2S to <0.1 ppm by volume). It is the preferred pathway for almost all units of process industry involving strict gas purification requirements [3]. By 2006, over 70 large enterprises saw different Rectisol units in their plants by patent and technology suppliers like Lurgi and Linde [4,5].

Although flourishing in industry, the description of thermodynamic behavior of the hypothermal process is rarely described, hindering accurate process simulation and further process melioration. According to preliminary trials, almost all possible thermodynamic models in prevailed commercial process simulators, such as ProII with Provision and ASPEN, were found to give no workable results compared with on-site data. In an effort to accurately simulate and improve a typical five-column scheme, we revised the thermodynamic parameters according to industrial data through regression analyses and predictions. The revisions of the thermodynamic model presented in this work perfectly describe the current process and facilitate a process optimization-oriented unit heat integration based on pinch and next step exergy analyses.

2. Simulation of the protytope rectisol process

2.1 Feed conditions

The feed gas in the simulation is from a carbon monoxide shift conversion device. The feed condition: flow rate 8160 Kmol/h, temperature 245.2K, pressure 3.63Mpa, The characteristics are listed in Table 1.

Table 1. Basic characteristics of the feed gas

Components	Concentration (mol%)
CH_3OH	5.44E-05
CO_2	0.3229
H_2	0.4576
N_2	2.81E-03
H_2S	9.95E-04
H_2O	3.59E-06
Ar	1.00E-03

CH_4	8.10E-04
CO	0.2138
COS	1.00E-15

2.2 Process description of the target Rectisol

The typical process of five-column Rectisol is shown schematically in Fig. 1. It is composed of five blocks: 1) acid gas absorption column (C1), 2) CO_2 regeneration column (C2), 3) H_2S concentration column (C3), 4) methanol regeneration column (C4) and 5) methanol/water separation column (C5).

The raw gas from the upstream carbon monoxide unit is cooled by hypothermal methanol in E1 and then flash liberated in V1 before flowing into C1. The moisture is frozen and solidified for removal, avoiding the adverse blocking effects in downstream equipment. Within this absorption operation, most H_2S and CO_2 are removed, leaving the purified raw gas which is transported to the liquid nitrogen wash unit for further treatment involving cryogenic recovery of carbon monoxide and oxo-gas. Two sidelines are drawn from the upper and lower column sections. The upper side cuts flow to C2, whereas CO_2 is flashed out before being transported to the urea or methanol synthesis unit and the lower to C3. The dissolved gas (except for H_2S) in the bottom product of the lower absorption column is stripped by low pressure N_2. In addition, the methanol solution, rich in H_2S, is preheated before being transported to C4. It is then regenerated through rectification, which is collaboratively powered by bottom reboiler heating and vapor stripping by the top product of C5. H_2S and other trace impurities are recovered at the top of the C4 column. Finally, the lean methanol is refined in C5 and pumped back to absorb raw material gas.

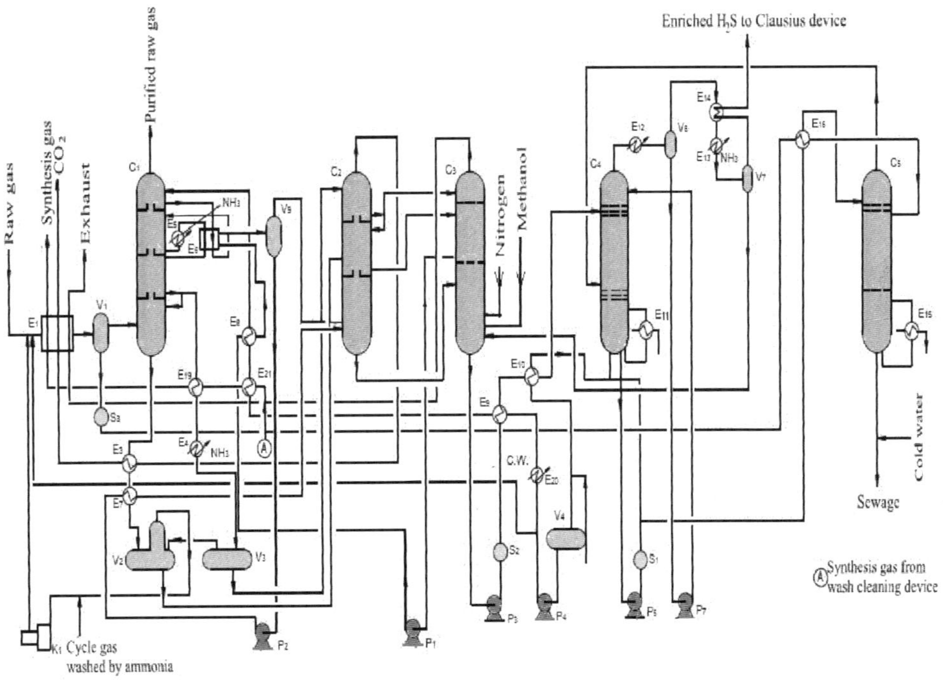

Figure 1. Flow sheet of five units Rectisol

2.3 Thermodynamic models

Occasionally, in process-oriented research, there is a specific, non-ideal material system or a special extreme operation condition under which the systemic thermodynamic behavior is far from normal. Similarly, the inbuilt thermodynamic models or simulation methods fail to accurately describe the absorptive process involving methanol, carbon dioxide and hydrogen sulfide under low temperature and high pressure conditions. In order to achieve a veracious simulative process description, revision of the thermodynamic model is inevitable. The convenient and fast approach involves selecting the thermodynamics models which are as close to on-site results as possible, then rectify the parameters of this thermodynamic prototype with field or experimental database.

Before the thermodynamic revision, we split the complex columns into functional blocks for simplification purpose [6, 7]. C1 is the first column in the process and deviates the most thermodynamically. This column was divided into 4 parts: 1) H_2S absorption section (C110), 2) CO_2 minor absorption section (C120), 3) main absorption section (C130) and 4) fine absorption section (C140). The next target, C3, falls into the upper column (C320) and the under column (C310).

Secondly, different models were screened in advance to simulate these blocks or sections suitably, including PSRK [8,9], RK-ASPEN, RKSWS, SR-POLAR, PRWS,
KSMHV2 and PRMHV2. As shown in Table 2, the appropriate thermodynamic models, whose simulative results most closely matched the field data, were adopted in the corresponding correction.

Table 2. Thermodynamics model of every block

Block	Model
C110	adjusted RK-ASPEN
C120	SR-POLAR
C130	SR-POLAR
C140	SR-POLAR
C200	RK-ASPEN
C310	PSRK
C320	RK-ASPEN
C400	PSRK
C500	PRWS

In C110, calculated results by RK-ASPEN illustrated that stage temperature and stream composition of most components, except for CO_2, resembled field data within permissible deviation. Considering this extreme situation, the authoritative data in the literature [10] was, for the first trial, adopted to modify the related parameters (tabulated in Table 3.) in RK-ASPEN model. Apparently, the simulative computation results supported by this modified model, listed in Table 4., verify a satisfactory effect in contrast with their industrial counterparts. Preliminary results are obtained using a previous RK-ASPEN model and final results were modified.

Table 3. Parameters regressed

Parameters	Component i	Component j	Valve
RKAPOL	CO_2		-0.6561249
RKAPOL	CH_4O		0.48309335
RKAKA0	CO_2	CH_4O	-0.4446355
RKAKA1	CO_2	CH_4O	1.74342955
RKAKB0	CO_2	CH_4O	-0.3010129
RKAKB1	CO_2	CH_4O	1.07304549

Table 4. Comparison between the simulation results and field data of steam103 in the C110 column

flow rate (Kmol/h)	field data	preliminary results	final results
CH_3OH	4.377	2.01218	1.7011
CO_2	2528.2345	1985.553	2536.7
H_2	3732.81	3741.108	3742.1
N_2	22.9006	23.09295	23.109
H_2S	0.00037	0.000327	0.0003
H_2O	0.000029	0.003208	0.0025
Ar	8.17532	8.142333	8.162
CH_4	6.5829	6.619009	6.6469
CO	1743.6932	1751.876	1753.3
COS	6.76E-12	3.61E-12	3.64E-12

3. Analysis and comparison of calculation results

According to the preparatory work above, the prototype flow sheet of the Rectisol process, shown in Fig. 2, was up built and simulated. Special attention was paid to comparisons between simulation results and field data, including stage temperature, total stream and main component flow rate.

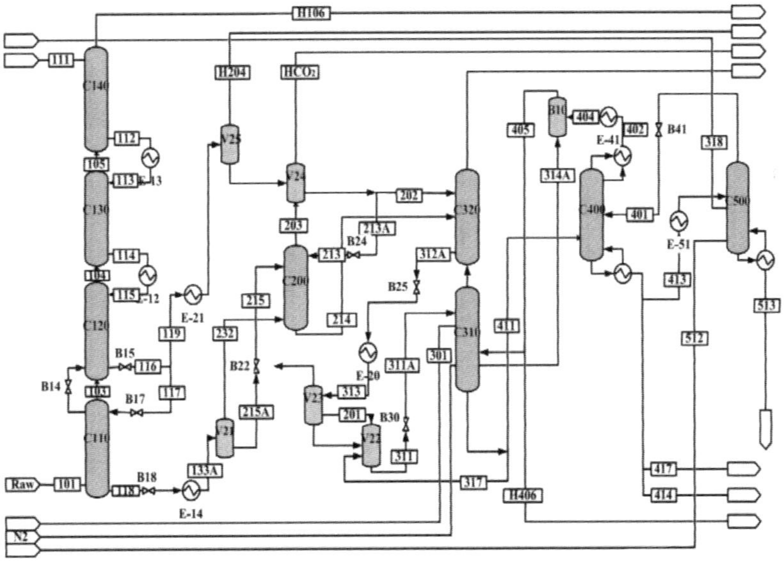

Figure 2. The base-case flow sheet of Rectisol process

Stream 103 in C110, as well as 104 in C120, 105 in C130, 106 in C140, 214 in C2, 312 and tail gas in C3, 401 in C400, 402 in C500, were chosen for this purpose.

Table 5. reveals that the simulation results agree well with the field data, with most errors less than 5%. For specified input feedstock and conditions, the results of simulation based on the revised thermodynamic models are reliable for process analysis and optimization. Their parameters were adjusted using a published database or regressive field correspondence. With continuous addition of more field data to the reservoir, the binary interaction parameters are expected to be more precise and versatile for various feeding conditions.

Table 5. Comparison between simulation and field data

Streams	Items	Simulation	Field data	RE/%
	CO_2 (Kmol/h)	2518.28	2528.24	0.39
103	T (K)	246.3	252.4	2.42
	F (Kmol /h)	8052.72	8046.78	0.07
	CO_2 (Kmol/h)	2016.91	2032.19	0.75
104	T (K)	245.3	249.3	1.6
	F (Kmol /h)	7549.63	7543.58	0.08
	CO_2 (Kmol/h)	1549.83	1548.99	0.05
105	T (K)	246.1	249.1	1.2
	F (Kmol /h)	7080.61	7053.99	0.38
	CO_2 (Kmol/h)	150.3	163.5	8.07
106	T (K)	221.8	232.5	4.6
	F (Kmol /h)	5669.44	5626.65	0.76
	CO_2 (Kmol/h)	318.27	318.57	0.09
203	T (K)	230.6	229.1	0.65
	F (Kmol /h)	325.21	325.43	0.07
	CO_2 (Kmol/h)	1153.06	1147.01	0.53
302	T (K)	258.4	240.3	7.53
	F (Kmol /h)	1605.5	1602.92	0.16
	CO_2 (Kmol/h)	2033.05	2020.8	0.61
Tail gas	T (K)	235.7	216.4	8.92
	F (Kmol /h)	2473.4	2471.01	0.1
	CH_3OH (Kmol/h)	2033.05	2020.8	0.61
311	T (K)	235.7	216.4	8.92
	F (Kmol /h)	2473.4	2471.01	0.1
	CH_3OH (Kmol/h)	103.34	104.67	1.27
401	T (K)	374.7	374.2	0.13
	F (Kmol /h)	107.67	107.68	0.01
	H_2S (Kmol/h)	45.25	45.67	0.92
402	T (K)	318.7	319.3	0.19
	F (Kmol /h)	139.87	139.88	0.01

4. Heat intergration and process optimization

The thermodynamic preparation facilitates systemic checking and analysis for flow sheet defects. The following revision or optimization will aid the energy-saving oriented exergy and enthalpy analyses. The concern here is the connective relationship of the blocks instead of the outer periphery blocks.

4.1 Heat integration

The original heat exchanger network (HEN) of the Rectisol process, as illustrated in Fig. 3 involved six sets of cold utility (CU) distributed in the six pre-cooling locations of five columns. CU is mainly offered through energy and a cost-intensive ammonia refrigeration system, which brings out the significance of more reasonable matching of heat sources and sinks in this scheme. With this in mind, necessary flow sheets were focused on the heat exchanging system, in hope of a more effective suggestion for energy conservation and absorption of acid gas.

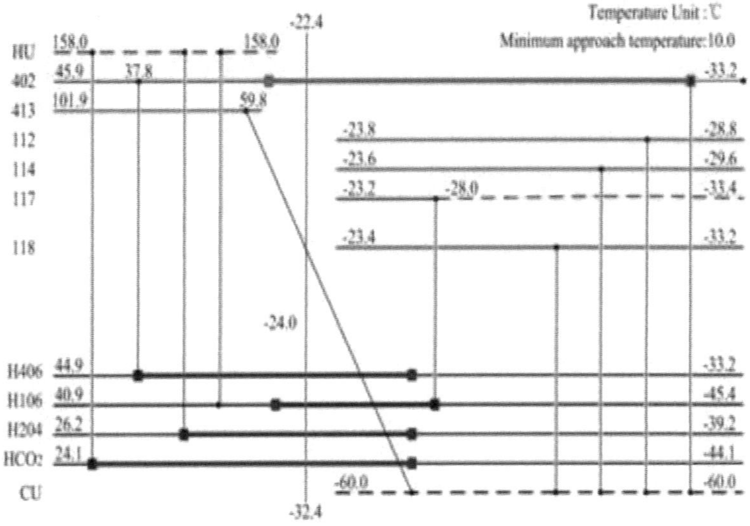

Figure 3. The original HEN

In order to reclaim the cryogenic energy within the system, four cold flows were added to the basis of the original HEN. These cold flows include the following: the separated gas from the overhead of the acid gas absorption column (H106), the flash tank top gas (H204), the Klaus gas (H402) and the flash tank top gas (labeled HCO2). The prefix, 'H', of these four cool streams separates them from the hot streams.

The original HEN was analyzed through pinch analysis [11, 12], using 10℃ as minimum approach temperature. This led to the following findings: 5 heat exchangers transfer heat across the pinch point, whereas the CU is used under the pinch point. These results are contrary to the principles and common senses of pinch analysis.

In order to obtain optimal HEN, the task was broken down into three steps. First, the appropriate minimum approach temperature was selected by thermodynamic and economic consideration. Fig. 4 shows the analysis of the economic condition at different minimum approach temperatures. With the minimum approach temperature increasing, the annual operating costs go up while the total heat transfer area decreases. Considering that the cost of heat-exchange equipment is lower than that of energy consumption, the largest area of heat transfer and the minimum approach temperature of 10℃ were finally adopted. Second, the temperature-enthalpy diagram of the original network and the total portfolio curve were plotted, as shown in Fig. 5 and Fig. 6. The streams, which stayed up and down the pinch, were re-matched in order to fully recover the heat and cold in the system. Finally, the optimal heat exchanger network could be obtained by appropriate adjustments, as shown in Fig. 7.

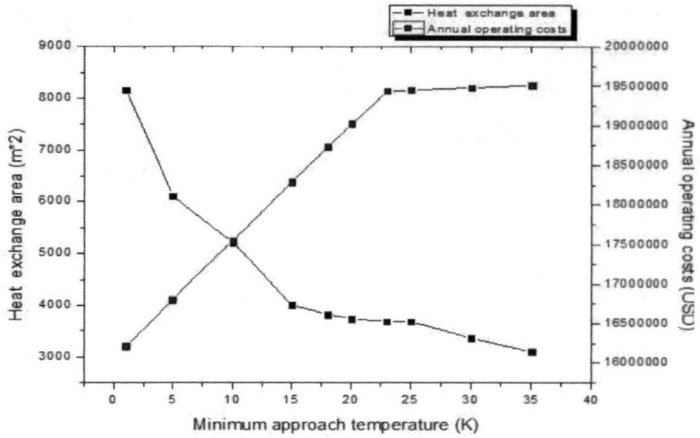

Figure 4. Variation of heat exchange area and annual operating costs due to minimum approach temperature

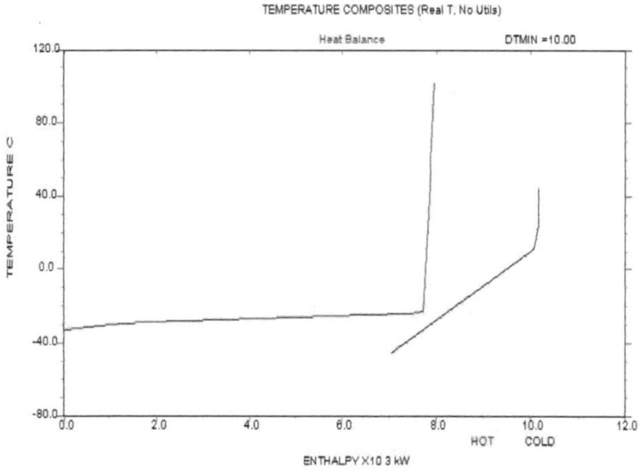

Figure 5. The temperature-enthalpy diagram of HEN at 10℃

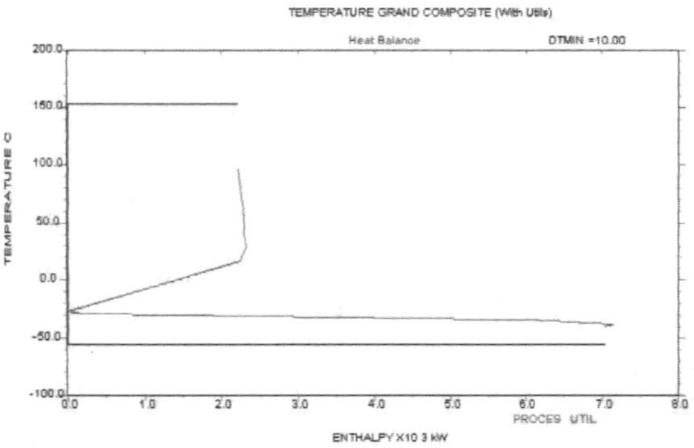

Figure 6. The GCC of HEN at 10℃

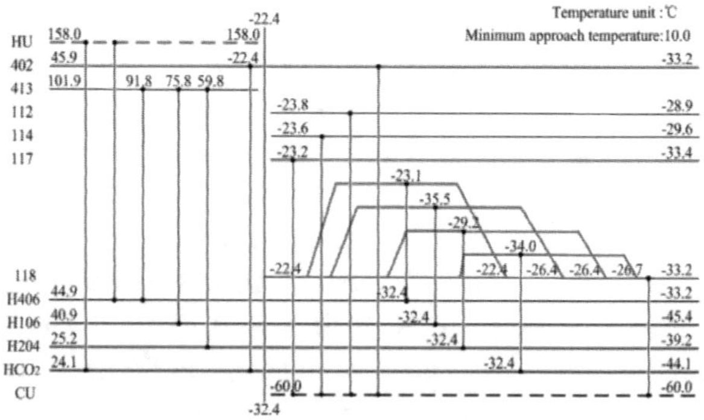

Figure 7. The optimized HEN

The optimized HEN was proven to be superior to the existing prototype network. At the cost of increasing the heat interchanging area of heat exchanger, 327.2 kw hot utilities (HU) and 356.4 kw cold utilities were calculably reclaimed. From an economic perspective, \$97,200 per year will be saved by merely adding one-off equipment and a construction investment worth \$32,800. It will take only 0.337years to recover the capital by advance estimation under specific stipulation of pinch analyses.

Compared with the original process, the new heat interchanging area increased by 36.36%, and HU and CU consumption were reduced 12.86% and 4.82%, respectively. Additionally, annual operating costs and annual aggregate costs were cut by 5.25% and 5.18%, in the new scheme. Detailed data is shown in the Table 6.

Table 6. Economic comparison between new heat exchanger network and original exchanger network

	Original HEN	New HEN
Number of heat exchanger	11	15
Heat transfer area (m^2)	3999.5	5213.3
Installation costs ($\times 10^3$\$)	104.8	137.6
Heat consumption (KW)	2544.9	2217.7
Cool consumption (KW)	7394.1	7037.7
Hot utility consumption ($\times 10^3$\$/Yr)	982. 6	856. 3
Cold utility consumption ($\times 10^3$\$/Yr)	17536.9	16691 .6
Annual capital cost ($\times 10^3$\$/Yr)	34 .9	45 .9
Annual operating costs ($\times 10^3$\$/Yr)	18519.5	17547. 9
Annual aggregate costs ($\times 10^3$\$/Yr)	18554.5	17593.8

4.2 *Process optimization*

The exergy efficiency of columns can be evaluated by the exergy analysis method [13] in the Rectisol process. The analytical results, shown in Table 7., indicate that the exergy efficiency of the methanol regeneration column and methanol/water separation column is not satisfactory. The reason is that external energy is added into the two columns. In the methanol regeneration column, cooling water is used for overhead condensing and product cooling. Similarly, the methanol/water separation column consumes external energy to evaporate the column bottom liquid. Therefore, simulative process rearranging is carried out in connection with two columns.

The transformation program of the methanol regeneration column is as follows: stream 411 is preheated from 88.5℃to 98.4℃by one tributary of stream 413, which is cooled from 101.94℃to 91.85℃. In addition, the side stream from the second tray is heated from 91.21℃ to 97 ~ 99℃ by the other tributary of stream 413, then returns to the previous tray. Those of methanol-water separation column are stream 511 and 318, which are heated from 57.9℃ and 55.3℃ to 71.75℃and 125℃, respectively, by wastewater from the bottom. Stream 512 is heated from -28℃ to 12.05℃ by circulating cooling water. After process optimization, annual hot utility consumption will be decreased by approximately 12505.33 kW, 11.5% of that of the unadjusted.

Table 7. Exergy efficiency of columns

Columns	Exergy loss	Exergy Input	Exergy efficiency
C1	4295.669	-2081377	99.894%
C2	1739.021	-452334.55	99.616%
C3	1939.537	-1251213.6	99.845%
C4	9185.14	-459841	98.003%
C5	78487.379	-503578	84.414%

5. Conclusion

Based on published reservoir and field data, a Rectisol process is numerically set up and simulated with the aid of a commercial process simulator. In other words, a secondary development of the program for a first-principles method is developed. Furthermore, HEN and process are optimized by the aid of two main thermodynamic analysis methods: pinch and exergy analysis. The main conclusions are as follows:

(1) The modified thermodynamic model is suitable for simulating the Rectisol process and favorably conformed to the industrial site measurement, with the accurate and reliable parameters regressively obtained.

(2) The original HEN was analyzed to have two unreasonable aspects. The problems were solved by the simulative re-matching streams and by adding four cold flows supported by the adjusted thermodynamic models.

(3) In all, \$97,200 per year will be saved at the price of \$32,800 one-off investment to improve HEN.

(4) The relatively low exergy efficiency of methanol regeneration column and methanol-water separation column is raised in the optimized system, with annual hot utility consumption decreased by 12505.33 kw.

(5) On the basis of this work, more can be done to improve industrial units or reconfigure more advanced schemes for high energy efficiency.

References

1. Weiss, H. (1988), "Rectisol wash for purification of partial oxidation gases" Gas Separation & Purification, pp.171-176.

2. Olajire, A.A.(2010), "CO2 capture and separation technologies for end-of-pipe applications – A review" Energy, pp. 2610-2628.

3. Sadhukhan, J. Zhao,Y.R. Shah, N. and Brandon, N.P. (2010), "Performance analysis of integrated biomass gasification fuel cell (BGFC) and biomass gasification combined cycle (BGCC) systems" Chemical Engineering Science, pp. 1942-1954.

4. Satoshi, K. Hideyuki, K. Tohru, S. and Kiyohisa, O. (2006), "Photoelectrochemical reduction of carbon dioxide at p-type gallium arsenide and p-type indium phosphide electrodes in methanol" Chem.Eng.J., pp. 227-231.

5. Kaneco, S. Iiba, KJ. Hiei, N.H. Kiyohisa, O. Mizuno, T. and Suzuki, T. (1999), " Electrochemical reduction of carbon dioxide to ethylene with high Faradaic efficiency at a Cu electrode in CsOH/methanol" Electrochimica Acta, pp. 4701-4706.

6. LIEBMANN, DHOLE K.V.R. and JOBSON, M. (1998), "Integrated Design of a Conventional Crude Oil Distillation Tower Using Pinch Analysis" Chemical Engineering Research and Design, pp.335-347.

7. Watzdorf R.V, Bausa, J. Marquardt, W. (1999), "Shortcut Methods for Nonideal Multicomponent Distillation: 2. Complex Columns" AIChE Journal, pp. 1615-1628.

8. Holderbaum, T. Gmehling, J. (1991), "PSRK: A Group Contribution Equation of State Based on UNIFAC" Fluid Phase Equilibria, pp. 251-265.

9. Fischer, K. Gmehling, J. (1995), "Further development, status and results of the PSRK method for the prediction of vapor-liquid equilibria and gas solubilities" Fluid Phase Equilibria, pp. 1-22.
10. Knapp, H. Döring, R. Oellrich, L. Plocker, U. Prausnitz, J.M. (1982), "Vapor-Liquid Equilibria for Low Boiling Substances" DECHEMA, Frankfurt, Binary Systems:part1.
11. Matijaseviæ, L. Otmaèiæ, H. (2002), " Energy recovery by pinch technology" Applied thermal Engineering, pp. 477-484.
12. Zhelev, T.K. Semkov K,A. (2004), " Cleaner flue gas and energy recovery through pinch analysis" Journal of Cleaner production, pp. 165-170.
13. Fawzi, B. Nesreen, J. (2008), " Exergy analysis of desalination by solar-powered membrane distillation units" Desalination, pp. 27-40.

() Divided Wall Column for Industrial Multi Purpose Use

Daniel Staak[1], Thomas Grutzner[2], Brian Schwegler[2]and Detlef Roederer[2], (1)Department of Process Dynamics and Operation, Technical University of Berlin, Berlin, Germany, (2)Lonza Chemical R&D, Visp, Switzerland

Divided wall columns (DWC) for the distillation of multicomponent mixtures have received much attention in the past 15 years and have experienced a booming development. Although the concept isn't new and originates from the 1930s [1], it took a couple of decades until scientific investigations focused again on heat coupled columns [e.g. 2, 3]. Finally BASF in Ludwigshafen set up the first production scale column in 1985 [4]. The main advantages of DWC are the considerable savings in energy consumption and investment costs. This became increasingly important in times of rising energy costs and increased market competition. Therefore the number of installed units continuously increased since the 80's and today this technique can be considered to be well established in chemical industry.

Since DWC demand higher efforts in equipment design and process control they are predominantly used in continuously operated dedicated production plants. Very little literature can be found where DWC are used in multi-purpose plants. The equipment in those plants, typically batch distillation, has to show a high degree of flexibility – an attribute that hardly seems to fit for divided wall columns.

Lonza set up a multi-purpose DWC on the production site in Visp that fully meets the demands of a steadily changing production. The column is constructed from hastelloy c22 which allows the distillation of corrosive mixtures used in this plant. A flexible inlet enables the feeding on different column segments as well as a flexible outlet allows a side draw from different stages. Also simultaneously feeding on both sides of the dividing wall is possible which allows an operation mode comparable to that of a traditional distillation column This paper reports on the equipment design and the specific technical solutions that had to be implemented in order to meet the harsh specifications of multi purpose production. Particular attention is put on the modelling and simulation of the column and the control scheme that has been applied. The launch and operation of the column is explained in detail and finally the original simulations are compared to real operation data.

[1] A. J. Brugma:*Process and Device for Fractional Distillation of Liquid Mixtures, more Particularly Petroleum, US Patent 2.295.2561942(filed 1936)*

[2] F. B. Petlyuk V. M. Platonov, D. M. Slavinskii:*Thermodynamically Optimal Method for Separating Multicomponent Mixtures*, Int. Chem. Eng. 5,**1965**, 555-561

[3] W. J. Stupin, F. J. Lockhart:*Thermally Coupled Distillation: a Case History*, Chem. Eng. Progr. 88,**1972**, 71-72

[4] BASF company website

AUTHOR INDEX